Communications in Microgrids

Communications in Microgrids

Special Issue Editors

**Peter Xiaoping Liu
Wenchao Meng
Hui Chen
Chuanlin Zhang**

MDPI • Basel • Beijing • Wuhan • Barcelona • Belgrade

Special Issue Editors

Peter Xiaoping Liu
Carleton University
Canada

Wenchao Meng
Zhejiang University
China

Hui Chen
Shanghai University of Electric Power
China

Chuanlin Zhang
Shanghai University of Electric Power
China

Editorial Office
MDPI
St. Alban-Anlage 66
4052 Basel, Switzerland

This is a reprint of articles from the Special Issue published online in the open access journal *Energies* (ISSN 1996-1073) in 2018 (available at: https://www.mdpi.com/si/energies/Communications_in_Microgrids).

For citation purposes, cite each article independently as indicated on the article page online and as indicated below:

LastName, A.A.; LastName, B.B.; LastName, C.C. Article Title. *Journal Name* **Year**, *Article Number*, Page Range.

ISBN 978-3-03928-482-5 (Pbk)
ISBN 978-3-03928-483-2 (PDF)

© 2020 by the authors. Articles in this book are Open Access and distributed under the Creative Commons Attribution (CC BY) license, which allows users to download, copy and build upon published articles, as long as the author and publisher are properly credited, which ensures maximum dissemination and a wider impact of our publications.

The book as a whole is distributed by MDPI under the terms and conditions of the Creative Commons license CC BY-NC-ND.

Contents

About the Special Issue Editors . . . . **vii**

Preface to "Communications in Microgrids" . . . . **ix**

Yueping Sun, Li Ma, Dean Zhao and Shihong Ding
A Compound Controller Design for a Buck Converter
Reprinted from: *Energies* **2018**, *11*, 2354, doi:10.3390/en11092354 . . . . **1**

Yantao Liao, Jun You, Jun Yang, Zuo Wang and Long Jin
Disturbance-Observer-Based Model Predictive Control for Battery Energy Storage System Modular Multilevel Converters
Reprinted from: *Energies* **2018**, *11*, 2285, doi:10.3390/en11092285 . . . . **18**

Mousa Marzband, Masoumeh Javadi, Mudathir Funsho Akorede, Radu Godina, Ameena Saad Al-Sumaiti, EdrisPouresmaeil
A Centralized Smart Decision-Making Hierarchical Interactive Architecture for Multiple Home Microgrids in Retail Electricity Market
Reprinted from: *Energies* **2018**, *11*, 3144, doi:10.3390/en11113144 . . . . **37**

Bo Tang, Gangfeng Gao, Xiangwu Xia and Xiu Yang
Integrated Energy System Configuration Optimization for Multi-Zone Heat-Supply Network Interaction
Reprinted from: *Energies* **2018**, *11*, 3052, doi:10.3390/en11113052 . . . . **59**

Shengnan Zhao, Beibei Wang, Yachao Li and Yang Li
Integrated Energy Transaction Mechanisms Based on Blockchain Technology
Reprinted from: *Energies* **2018**, *11*, 2412, doi:10.3390/en11092412 . . . . **77**

About the Special Issue Editors

Peter Xiaoping Liu received his B.Sc. and M.Sc. degrees from Northern (Beijing) Jiaotong University, Beijing, China in 1992 and 1995, respectively, and Ph.D. degree from University of Alberta, Edmonton, AB, Canada in 2002. He has been with the Department of Systems and Computer Engineering, Carleton University, Ottawa, ON, Canada since 2002, where he is currently a Professor and a Canada Research Professor. His current research interests include teleoperation, haptics, surgical simulation, and control systems. Dr. Liu has served as an Associate Editor for several journals, including *IEEE/ASME Transactions on Mechatronics*, *IEEE Transactions on Cybernetics*, *IEEE Transactions on Automation Science and Engineering*, and *IEEE Transactions on Instrumentation and Measurement*. He is a Licensed Member of the Professional Engineers of Ontario (P. Eng.), a fellow of the Institute of Electrical and Electronics Engineers (FIEEE) and a fellow of the Engineering Institute of Canada (FEIC).

Wenchao Meng received the Ph.D. degree in control science and engineering from Zhejiang University, Hangzhou, China, in 2015. He is currently a Professor with Zhejiang University, Hangzhou, China. His current research interests include adaptive intelligent control, cyber-physical systems, renewable energy systems, and smart grids.

Hui Chen received her B.S. degree in Control and Instrument Specialty from Jiangsu University, P.R. China, in 2002, the M.S. and Ph.D. degree in Control Science and Engineering at Shanghai University in 2006. She was a Visiting Ph.D. Student with the Department of IRSEEM, ESIGELEC Rouen, France, from 2010 to 2011. She was a joint PhD in Computation department of Jacobs University in Bremen from December 2011 to December 2012. She was a Visiting Scholar with the Computer Science Institute, Curtin University, Australia, from 2017 to 2018. Currently, she is a lecturer of Shanghai University and Electric Power at the College of Automation Engineering with interests in pattern recognition, computer vision, and deep learning.

Chuanlin Zhang (Professor) received the B.S. degree in mathematics and the Ph.D. degree in control theory and control engineering from the School of Automation, Southeast University, Nanjing, China, in 2008 and 2014, respectively. He was a Visiting Ph.D. Student with the Department of Electrical and Computer Engineering, University of Texas at San Antonio, USA, from 2011 to 2012; a Visiting Scholar with the Energy Research Institute, Nanyang Technological University, Singapore, from 2016 to 2017; a visiting scholar with Advanced Robotics Center, National University of Singapore, from 2017 to 2018. Since 2014, he has been with the College of Automation Engineering, Shanghai University of Electric Power, China, where he is currently a Professor of Special Appointment (Eastern Scholar) at Shanghai Institute of Higher Learning. He is the principal investigator of several research projects, including Eastern Scholar Program at Shanghai, Leading Talent Program of Shanghai Science and Technology Commission, Chenguang Program by the Shanghai Municipal Education Commission, etc. His research interests include nonlinear system control theory and applications for power systems.

Preface to "Communications in Microgrids"

The microgrid, as a small-scale power system, is expected to continue to grow with smartness, providing increased reliability and facilitating effective integration of distributed generators and energy storage devices. These new capabilities are made possible by integrating advanced control methods (e.g., model predictive control, nonlinear control) and advanced communication technologies (e.g., home area networks, field area networks and wide area networks). The combination of advanced control methods and advanced communication technologies are key enablers for various future microgrid applications, such as demand response, advanced metering infrastructure (AMI), energy storage integration, electric vehicle (EV) charging and seamless integration of renewable energy sources.

This book introduces the advanced control and communication methods for microgrids, which includes 5 chapters. In chapter 1, a compound controller is designed for a buck converter based on disturbance observer (DO) and nonsingular terminal sliding mode (TSM). In chapter 2, a model predictive control is studied for a battery energy storage system based on a disturbance observer. In chapter 3, a combined market operator and a distribution network operator structure have been devised for multiple home-microgrids connected to an upstream grid. In chapter 4, optimization of an integrated energy system aimed at improving the comprehensive utilization of energy through cascade utilization and coordinated scheduling of various types of energy is studied. In chapter 5, a distributed integrated energy transaction mechanism for power market based on the blockchain technology is proposed.

Peter Xiaoping Liu, Wenchao Meng, Hui Chen, Chuanlin Zhang
Special Issue Editors

Article

A Compound Controller Design for a Buck Converter

Yueping Sun [1,2], Li Ma [1,*], Dean Zhao [1,2] and Shihong Ding [1]

1. School of Electrical and Information Engineering, Jiangsu University, Zhenjiang 212013, China; sunypujs@mail.ujs.edu.cn (Y.S.); dazhao@mail.ujs.edu.cn (D.Z.); dsh@mail.ujs.edu.cn (S.D.)
2. School Key Laboratory of Facility Agriculture Measurement and Control Technology and Equipment of Machinery Industry, Jiangsu University, Zhenjiang 212013, China
* Correspondence: mali@mail.ujs.edu.cn; Tel.: +86-511-8879-1245

Received: 24 July 2018; Accepted: 4 September 2018; Published: 6 September 2018

Abstract: In order to improve the performance of the closed-loop Buck converter control system, a compound control scheme based on nonlinear disturbance observer (DO) and nonsingular terminal sliding mode (TSM) was developed to control the Buck converter. The control design includes two steps. First of all, without considering the dynamic and steady-state performances, a baseline terminal sliding mode controller was designed based on the average model of the Buck converter, such that the desired value of output voltage could be tracked. Secondly, a nonlinear DO was designed, which yields an estimated value as the feedforward term to compensate the lumped disturbance. The compound controller was composed of the terminal sliding mode controller as the state feedback and the estimated value as the feedforward term. Simulation analysis and experimental verifications showed that compared with the traditional proportional integral derivative (PID) and terminal sliding mode state feedback control, the proposed compound control method can provide faster convergence performance and higher voltage output quality for the closed-loop system of the Buck converter.

Keywords: terminal sliding mode; DC-DC converter; disturbance observer

1. Introduction

Switching power supplies are power conversion devices that provide the required voltage or current through different architectures. Although widely used in various fields, the control performances are not satisfactory under some large disturbance signals [1–4]. This is because most of them are based on proportional integral derivative (PID) control methods, while PID controllers may not overcome the adverse effect of large disturbance signals [5,6]. To this end, many scholars have devoted themselves to researching nonlinear controller designs for DC-DC converters, such as sliding mode control [7–9], fuzzy control [10–12], neural network [13,14], and intelligent control [15–17]. Among them, sliding mode control has been found to be one of the most effective methods to handle nonlinear uncertain systems, since sliding mode control is insensitive to system uncertainties, external disturbances, and parameter perturbations [18]. Consequently, sliding mode control has been applied to many practical systems, such as motors, power systems, robots, spacecraft, and servo systems [19–21].

Recently, sliding mode has also been applied to the control of DC-DC converters. For example, a method to implement a global switching function in a sliding mode controller was reported in Reference [22] for the first time, where the Buck converter's steady-state operation and output voltage ripple was analyzed and the transient condition criteria of the global closed-loop sliding mode control system was proposed. Compared with traditional sliding mode control, the sliding mode method proposed in Reference [22] exhibits faster transient load characteristics and better robustness. Also, the authors of Reference [23] proposed a method of design for a proportional-integral-like sliding mode controller, which uses an adaptive controller to compensate the error caused by the load fluctuation, thereby reducing the system's steady-state error. Meanwhile, the sliding mode controller proposed in Reference [23] also improves the steady-state and dynamic performance of the converter and facilitates the optimization of the controller parameters. Additionally, an adaptive terminal sliding mode (TSM) control strategy was proposed in Reference [24], which ensures that the output voltage error converges to the equilibrium point within a finite time. Furthermore, the adaptive law in Reference [24] can be integrated into the terminal sliding mode control strategy to achieve dynamic sliding during load fluctuation so as to improve the accuracy of system tracking. The authors of Reference [25] proposed a novel nonsingular terminal sliding mode manifold incorporating a disturbance estimation technique subject to matched/mismatched resistance load disturbances, and the proposed controller was found to improve tracking performance and disturbance rejection ability against resistance load variation.

Although there are many sliding mode control results for DC-DC converters, most of them are pure state feedbacks [26,27]. This implies that when the lumped disturbances are large, the only method to improve the tracking accuracy is to tune the sliding mode controllers' gains. It is known that the high-gain state feedback usually brings some shortcomings, such as a large overshoot, exciting unmodeled dynamics, and even instability [28,29]. Meanwhile, the high gains also bring the chattering problem [30]. This is because the chattering is usually proportional to the magnitude of the discontinuous terms, while the high gains are always the parameters of these discontinuous terms.

To resolve the above problem, the idea of a compound controller was developed in this paper to improve the performance of the DC-DC Buck converter's control system. By a combination of the nonsingular terminal sliding mode technique and the disturbance observer (DO) design method, a compound control scheme was developed step by step. The terminal sliding mode controller was designed to improve the disturbance rejection property, while the disturbance observer was constructed to further improve the dynamic performance of the closed-loop system. By comparing with the conventional terminal sliding mode and PID control schemes, the proposed compound algorithm was confirmed to provide a better dynamic and steady-state performance.

2. Problem Description

The circuit diagram of a Buck converter is shown in Figure 1, consisting of a DC voltage source, a switch tube SD, a diode D, an inductor L, a capacitor C, and a load resistor R_L. i_L is the inductive current, u_c is the output voltage, and U_s is the source voltage.

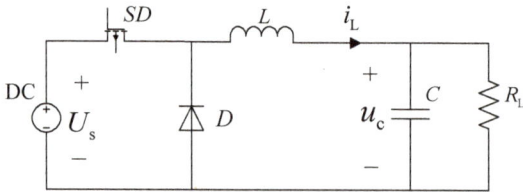

Figure 1. Circuit diagram of the Buck converter.

Since the switch has two states of "on" and "off", the Buck converter also has two working modes. According to the two different conditions, the average state model of the Buck converter can be established as:

$$\begin{cases} \frac{di_L}{dt} = \frac{1}{L}(\kappa U_s - u_c) \\ \frac{du_c}{dt} = \frac{1}{C}(i_L - \frac{u_c}{R_L}) \end{cases} \quad (1)$$

where κ represents the switch state, which is 1 for the "on" state of the switch and 0 for the "off" state.

Furthermore, considering the effect of disturbance on the system modeling [31–33], the above expression can be written as:

$$\begin{cases} \dot{i}_L = \frac{\kappa(U_s + \Delta U_s) - u_c}{L + \Delta L} + d_e(t) \\ \dot{u}_c = \frac{i_L - u_c/(R_L + \Delta R_L)}{C + \Delta C} \end{cases} \quad (2)$$

where the parameters ΔU_s, ΔL, ΔR_L, and ΔC are parameter perturbations, while $d_e(t)$ represents the corresponding system uncertainty and external disturbance. It is assumed that $d_e(t)$ and $\dot{d}_e(t)$ are both bounded, and then Equation (2) can be transformed to the following form:

$$\begin{cases} \dot{i}_L = \frac{\kappa U_s - u_c}{L} + \zeta_1(t) \\ \dot{u}_c = \frac{i_L - u_c/R_L}{C} + \zeta_2(t) \end{cases} \quad (3)$$

where $\zeta_1(t)$ and $\zeta_2(t)$ are expressed as:

$$\zeta_1(t) = \frac{\kappa \Delta U_s L - \kappa \Delta L U_s + \Delta L u_c}{(L + \Delta L)L} + d_e(t) \quad (4)$$

$$\zeta_2(t) = \frac{u_c \Delta R_L}{R_L(R_L + \Delta R_L)(C + \Delta C)} + \frac{u_c \Delta C - i_L \Delta C R_L}{CR_L(C + \Delta C)} \quad (5)$$

Since $d_e(t)$, ΔU_s, ΔL, ΔR_L, and ΔC are all bounded, this implies that $\zeta_1(t)$ and $\zeta_2(t)$ are also bounded.

The control objective of this paper is to design a compound control scheme based on nonsingular TSM and nonlinear DO for the Buck converter, so that the desired value of the output voltage of the system can be quickly tracked under disturbance.

3. Compound Controller Design

3.1. Nonsingular Terminal Sliding Mode Controller

We set the output voltage error to be $e_1 = u_c - V_0$, where V_0 is the DC reference output voltage. Based on Equation (3), the system's error dynamics can be expressed as:

$$\begin{cases} \dot{e}_1 = e_2 \\ \dot{e}_2 = \kappa \frac{U_s}{CL} - \frac{V_0}{CL} - \frac{e_1}{CL} - \frac{e_2}{CR_L} + \zeta(t) \end{cases} \quad (6)$$

where the system disturbance $\zeta(t)$ includes both $\zeta_1(t)$ and $\zeta_2(t)$, and can be expressed as:

$$\zeta(t) = \frac{\zeta_1(t)}{C} - \frac{\zeta_2(t)}{CR_L} + \dot{\zeta}_2(t) \quad (7)$$

Since $d_e(t)$ and its derivative are both bounded, from Equations (4) and (5), it is known that there exists constant C_λ and C_δ to make:

$$|\zeta(t)| \leq C_\lambda, |\dot{\zeta}(t)| \leq C_\delta \quad (8)$$

Let $\kappa \frac{U_s}{CL} = g(e)$ and $f(e) = \frac{V_0}{CL} + \frac{e_1}{CL} + \frac{e_2}{CR_L}$.
Then, System (6) can be expressed as:

$$\begin{cases} \dot{e}_1 = e_2 \\ \dot{e}_2 = g(e)u - f(e) + \zeta(t) \end{cases} \quad (9)$$

We designed the nonsingular terminal sliding mode surface as:

$$s = e_1 + \frac{1}{\alpha} e_2^{\frac{m}{n}} \tag{10}$$

where m and n are odd integers, and α, m, and n satisfy: $\alpha > 0$, $m > n > 0$, $1 < \frac{m}{n} < 2$.

The nonsingular terminal sliding mode controller was designed as:

$$u = g^{-1}(e)(f(e) - \frac{\alpha n}{m} e_2^{2-\frac{m}{n}} - \mu \cdot \text{sign}(s)) \tag{11}$$

where $\mu > C_\lambda + \eta, \eta > 0$, η is any real number. It can be verified that under Controller (11), the sliding variable will converge to the origin in a finite time.

The stability analysis of the finite-time convergence of closed-loop Systems (9) and (11) is given as follows.

Combined with Equation (9), the derivative of the sliding surface s is:

$$\dot{s} = \dot{e}_1 + \frac{m}{\alpha n} e_2^{\frac{m}{n}-1} \dot{e}_2 = e_2 + \frac{m}{\alpha n} e_2^{\frac{m}{n}-1}(g(e)u - f(e) + \zeta(t)) \tag{12}$$

Substituting Controller (11) into Equation (12) yields:

$$\dot{s} = -\frac{m}{\alpha n} e_2^{\frac{m}{n}-1}(-\mu \cdot \text{sign}(s) + \zeta(t)) \tag{13}$$

It is clear from Equation (13) that:

$$s\dot{s} = -\frac{m}{\alpha n} e_2^{\frac{m}{n}-1}(\mu|s| - \zeta(t)s) \leq -\frac{m}{\alpha n} e_2^{\frac{m}{n}-1} |s|(\mu - |\zeta(t)|) \tag{14}$$

With $|\zeta(t)| \leq C_\lambda$ and $\mu > C_\lambda + \eta$ in mind, we obtained:

$$s\dot{s} \leq -\frac{m\eta}{\alpha n} e_2^{\frac{m}{n}-1} |s| \tag{15}$$

Next, we needed to prove that under Controller (11), the state of System (9) will converge to zero within a finite time. On the one hand, it is easy to know that $e_2^{\frac{m}{n}-1} > 0$ when the system state $e_2 \neq 0$.

According to the finite-time Lyapunov theorem [34–37], the system state will converge to zero within a finite time. On the other hand, when the system trajectories stay in the line $e_2 = 0$, substituting Controller (11) into System (9) produces the following:

$$\dot{e}_2 = -\frac{\alpha n}{m} e_2^{2-\frac{m}{n}} - \mu \cdot \text{sign}(s) + \zeta(t) \tag{16}$$

which implies that when $e_2 = 0$, there is:

$$\dot{e}_2 = -\mu \cdot \text{sign}(s) + \zeta(t) \tag{17}$$

It is clear from Equation (17) that $s > 0$ when $\dot{e}_2 < 0$; conversely, $s < 0$ when $\dot{e}_2 > 0$. This means that the trajectory of the system will not stay on the axis $e_2 = 0$.

In conclusion, under Controller (11), the state of System (9) will converge to zero within a finite time.

Controller (11) is discontinuous and has severe chattering problems. In this paper, we employed the boundary layer method to eliminate the chattering, and thus the nonsingular terminal sliding mode Controller (11) can be rewritten as:

$$u = g^{-1}(e)(f(e) - \frac{\alpha n}{m} e_2^{2-\frac{m}{n}} - \mu \cdot \text{sat}(s)) \tag{18}$$

where the saturation function $sat(s)$ can be defined as: $sat(s) = \begin{cases} \varepsilon \, \text{sign}(s), & |s| > \varepsilon \\ s, & |s| \leq \varepsilon \end{cases}, \forall \varepsilon > 0.$

3.2. Nonlinear Disturbance Observer Design

Consider the following nonlinear system:

$$\begin{cases} \dot{x} = F(x) + G_1(x)u + G_2(x)D \\ y = f(x) \end{cases} \tag{19}$$

where x, u, D, y are the system state, system input, disturbance, and system output respectively; $F(x)$, $G_1(x)$, $G_2(x)$, and $f(x)$ are known functions.

According to the theory of disturbance observer [38], the nonlinear disturbance observer is designed as:

$$\begin{cases} \dot{P} = -L'G_2(x)P - L'[G_2(x)L's + F(x) + G_1(x)u] \\ \hat{D} = P + L's \end{cases} \tag{20}$$

where P is an internal state of the nonlinear DO. By combining Equation (20) and Buck converter's sliding mode control system model, expressed by System (12), the following disturbance observer can be designed:

$$\begin{cases} \dot{P} = -L'\frac{m}{an}e_2^{\frac{m}{n}-1}P - L'[\frac{m}{an}e_2^{\frac{m}{n}-1}L's + e_2\frac{m}{an}e_2^{\frac{m}{n}-1}f(e) + \frac{m}{an}e_2^{\frac{m}{n}-1}g(e)u] \\ \hat{D} = P + L's \end{cases} \tag{21}$$

The stability of the above disturbance observer is given as follows.
Letting $\tilde{D} = D - \hat{D}$, the derivative of the disturbance deviation is:

$$\dot{\tilde{D}} = \dot{D} - \dot{\hat{D}} = \dot{D} - (\dot{P} + L'\dot{s}) \tag{22}$$

Substituting (18) and (21) into Equation (22), the following can be obtained:

$$\dot{\tilde{D}} = \dot{D} + L'\frac{m}{an}e_2^{\frac{m}{n}-1}P + L'^2\frac{m}{an}e_2^{\frac{m}{n}-1}s - L'\frac{m}{an}e_2^{\frac{m}{n}-1}D(t) = \dot{D} - L'\frac{m}{an}e_2^{\frac{m}{n}-1}\tilde{D} \tag{23}$$

Select a Lyapunov function as:

$$V = \frac{1}{2}\tilde{D}^2 \tag{24}$$

Taking a derivative of $V = \frac{1}{2}\tilde{D}^2$ along System (23) yields:

$$\dot{V} = \tilde{D}\dot{\tilde{D}} = \tilde{D}\dot{D} - L'\frac{m}{an}e_2^{\frac{m}{n}-1}\tilde{D}^2 \tag{25}$$

From Equation (8), it is clear that $|\dot{D}(t)| \leq C_\delta$, which indicates that:

$$\dot{V} = \tilde{D}\dot{\tilde{D}} = -L'\frac{m}{an}e_2^{\frac{m}{n}-1}\tilde{D}^2 + C_\delta|\tilde{D}| \tag{26}$$

It can be easily verified that the disturbance error will converge to a small area of the origin.

In conclusion, a compound controller obtained by combining the terminal sliding mode state feedback (Equation (18)) and disturbance observer (Equation (21)) can be constructed as follows:

$$u = g^{-1}(e)(f(e) - \frac{an}{m}e_2^{2-\frac{m}{n}} - \mu sat(s) - \hat{D}) \tag{27}$$

Remark 3.1: From a theoretical point of view, the boundary level should be as small as possible. However, the small saturation level may cause chattering problems. Hence, the choice of the boundary level is a trade-off. For the disturbance observer, it can be seen from Equation (23) that a larger L' implies a smaller observation error. However, it is interesting that when we tune the parameter L' to be large enough, the performance of the observation will be unchanged. This may be caused by the hardware.

The block diagram of the compound controller for the Buck converter is shown in Figure 2, where the output voltage and inductive current information can be obtained from sensors, and the control output will generate a pulse width modulation (PWM) signal.

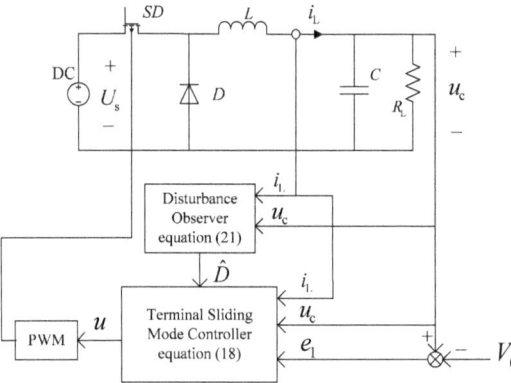

Figure 2. Block diagram of the compound controller for the Buck converter.

4. Simulation Analysis

To verify the feasibility and effectiveness of the proposed algorithm, MATLAB simulations were performed under three kinds of disturbance: start-up, step-load, and step-input-voltage. The converter parameters are given in Table 1.

Table 1. Parameters of the converter.

Parameter	Value
Input voltage, U_s	30 V
Inductance, L	330 µH
Capacitance, C	1000 pF
Load resistance, R_L	25 Ω
Voltage reference, V_0	15 V

In order to show the advantages of the proposed algorithm, the traditional PID control, terminal sliding mode control (TSM) and compound control (TSM + disturbance observer (DOB)) methods were compared. Firstly, the controller parameters were tuned so that the system under each controller could obtain the best convergence performance. The criterion was to tune the parameters to achieve the fastest convergence without considering the disturbance rejection property. Based on this, the PID parameters were taken as $K_p = 8$, $K_i = 5$, and $K_d = 0.2$, while the parameters of Controllers (18) and (27) were set as $\alpha = 3$, $m = 9$, and $n = 7$. The boundary layer level was set as $\varepsilon = 0.5$ and the disturbance observer parameter was chosen to be $L' = 40$. For comparison, the simulated start-up waveforms of the traditional PID control, terminal sliding mode (TSM) control, and the compound control (TSM + DOB) methods are shown in Figure 3. The convergence times for the first two cases were both about 0.4 s, while the compound control was found to converge to zero much more quickly—within 0.1 s.

For the disturbance rejection property, the PID controller was always worse than the TSM and compound controllers. This is because the steady-state error under TSM and the compound controllers can be steered to the origin in a finite time, while there always exists a steady-state error under the PID controller. This also implies that no matter what values of parameters are selected, the disturbance rejection properties of the TSM and compound controllers in the simulation will always be better than that of the PID controller.

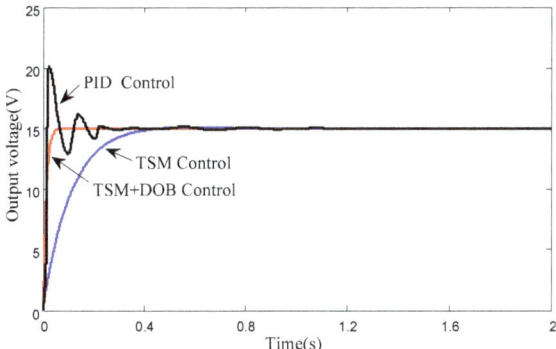

Figure 3. Simulated start-up waveform of output voltage in the absence of disturbance.

As a matter of fact, from a theoretical point of view, the steady-state error under TSM could be zero in a finite time, while the steady-state error under PID will always be restricted in the region of origin. It is apparent that from the theoretical point of view that the TSM controller could provide a better disturbance rejection property than the PID controller. Hence, we omitted the simulation performed under PID. In Figure 4, the load resistance steps from 25 Ω to 500 Ω at $t = 1$ s, and back to 25 Ω at $t = 1.5$ s. The output voltage has a response similar to that of the load resistance. Under the compound controller, it can be observed that the output voltage can return to the steady-state value quickly, and its convergence speed is obviously faster than that of the terminal sliding mode (TSM) controller. The response curves of the inductive current to the step-load are shown in Figure 5. It can also be seen that the current under the compound control can return back to its steady-state value quickly. Therefore, one can conclude that the compound controller with the disturbance observer can provide the system with a faster response speed and better disturbance rejection performance.

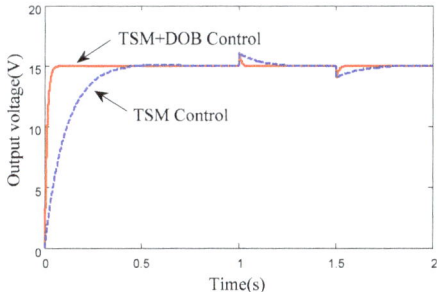

Figure 4. Simulated step-load waveform of the output voltage.

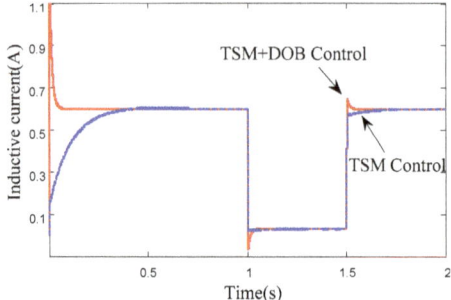

Figure 5. Simulated step-load waveform of the inductive current.

The simulated waveform of the output voltage and inductive current with respect to the input voltage stepping from 30 V to 40 V at $t = 1$ s and back to 30 V at $t = 1.5$ s are shown in Figures 6 and 7, respectively. Figure 6 shows that the value of the output voltage increases with the rise of the input voltage. Compared with the traditional terminal sliding mode control, it can be observed that the compound controller with the disturbance observer makes a smaller amplitude change and can quickly converge to the desired value. From Figure 7, we can see that the inductive current under both controllers exhibits a sudden change under TSM and TSM + DOB controllers when the input voltage changes. Nevertheless, the inductive current under the compound controller will reach the steady state rapidly, while the current under the traditional terminal sliding mode controller needs a period of recovery time to reach its steady-state value. In summary, the compound controller has a better control performance.

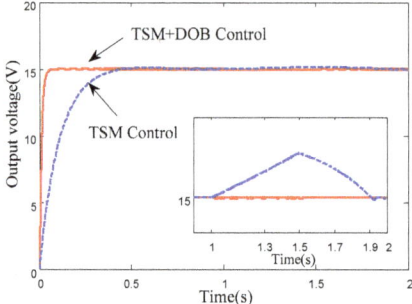

Figure 6. Simulated step-input-voltage waveform of the output voltage.

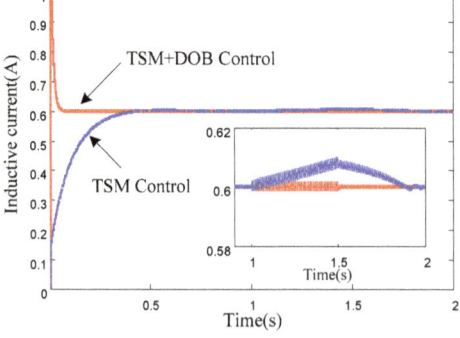

Figure 7. Simulated step-input-voltage waveform of the inductive current.

5. Experimental Verification

The circuit used in this experiment, shown in Figure 8, was the main circuit of a Buck converter with a 30-V DC voltage as the input. The control algorithms were implemented using digital signal processing (DSP) TMS320F28335 (Texas Instruments Inc., Dallas, TX, USA) with a clock frequency of 150 MHz. The voltage detection adopted the method of parallel resistance, which connects the two series resistors in parallel and adjusts their proportional relationship to meet the voltage sampling range of 0–3.3 V for DSP. The inductive current was measured using the ACS712 current module (Allegro MicroSystems LLC, Worcester, CM, USA). The analog signals of output voltage and inductive current were converted to digital signals through two 12-b analog-to-digital (A/D) converters. The resolution of digital pulse width modulation (DPWM) was 16 bits. The drive circuit adopted TLP250 (Toshiba Inc., Minato-ku, Tokyo, Japan) produced by Toshiba, and the PWM output of DSP was taken as its input signal. Meanwhile the IR2110 chip (International Rectifier Inc., Los Angeles, SC, USA) was bootstrapped, so that the PWM output amplitude was enough to operate the switch. The schematic diagram of the hardware is shown in Figure 8. The TLP250 provided both isolation and driving. In order to stabilize its built-in high-gain amplifier, a small ceramic capacitor and two current-limiting resistors must be placed between b1 and b3. The parameters of the components depend on the operating current of the luminous diode in the chip.

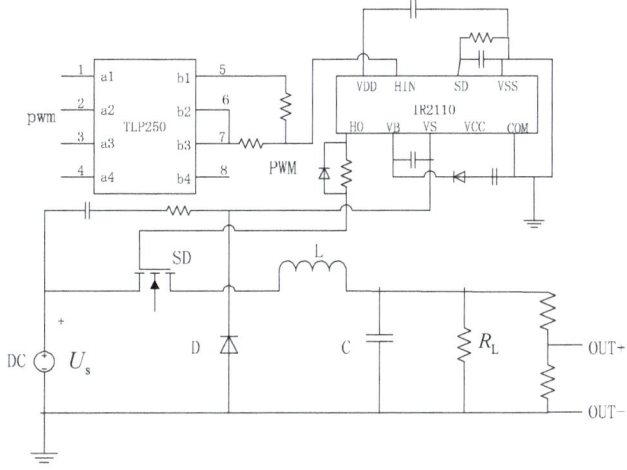

Figure 8. The schematic diagram of the hardware.

The software of this experiment used DSP as the control chip of the control loop. DSP is widely used in various fields of power electronics because of its fast execution speed, high efficiency, multi-function, and real-time control. The block diagram of the experimental platform is shown in Figure 9. The experimental setup is shown in Figure 10.

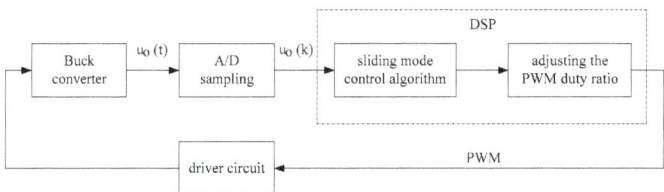

Figure 9. The block diagram of the experimental platform.

Figure 10. Experimental setup.

The parameters of the electronic elements are: $L = 330$ µH, $C = 1000$ pF, $R_L = 50$ Ω. The TSM coefficients are: $\alpha = 3$, $m = 5$, $n = 3$, while the PID control parameters are: $K_p = 10$, $K_i = 5$, $K_d = 0.1$.

When the input voltage is 30 V and the reference value of the input voltage is 15 V, experimental DC coupled start-up waveforms of the output voltage and inductive current under three control schemes (PID, TSM, TSM + DOB) are shown in Figures 11–13. The experimental start-up comparisons of the output voltage are given in Figure 14.

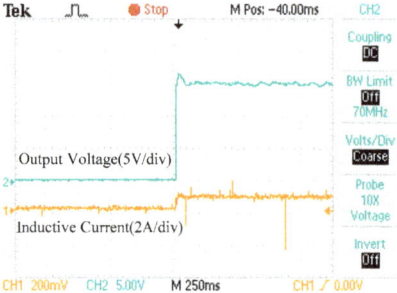

Figure 11. Experimental start-up waveforms of the output voltage and inductive current under the proportional integral derivative (PID) controller.

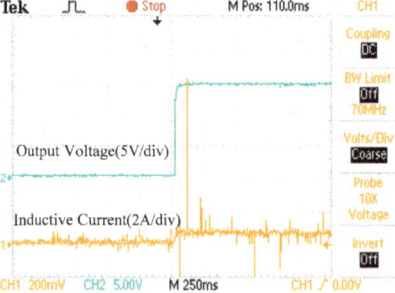

Figure 12. Experimental start-up waveforms of the output voltage and inductive current under the terminal sliding mode (TSM) controller.

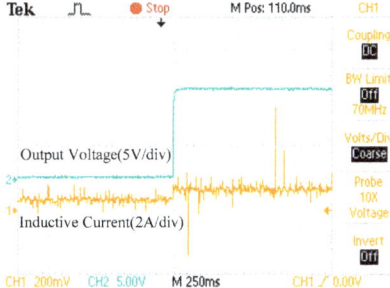

Figure 13. Experimental start-up waveforms of the output voltage and inductive current under the TSM + disturbance observer (DOB) controller.

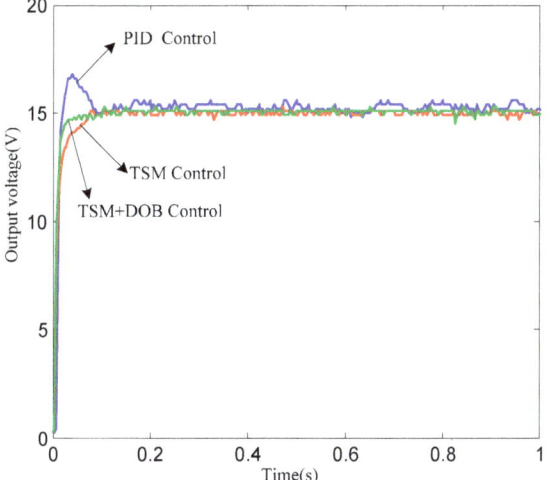

Figure 14. Experimental start-up comparisons of the output voltage under PID, TSM, and TSM + DOB controllers.

Table 2 gives the comparisons of overshoot, rising time, and settling time under PID, TSM, and TSM + DOB controllers at the start-up. It can be seen that the overshoot under the TSM + DOB controller is smaller than that under the other two control modes. The rising time under the TSM + DOB controller is shorter than that under the other two control modes, with the related rates of 36.4% and 145.5% when comparing with the PID and TSM control modes, respectively. In addition, the settling time under the TSM + DOB controller is also smaller than that under the PID and TSM control modes, with the related rates of 154.5% and 54.5%, respectively.

Table 2. Comparisons under proportional integral derivative (PID), terminal sliding mode control (TSM), and TSM + disturbance observer (DOB) at the start-up.

Controller	Overshoot (V)	Rising Time (ms)	Settling Time (ms)
PID	1.8	41.7	388.9
TSM	0.4	75.0	236.1
TSM + DOB	0.3	30.6	152.8

Experimental DC coupled step-load waveforms of the output voltage and inductive current under three control schemes (PID, TSM, TSM + DOB) when the load steps from 50 Ω to 100 Ω are shown in Figures 15–17, and the experimental step-load comparisons of the output voltage are given in Figure 18.

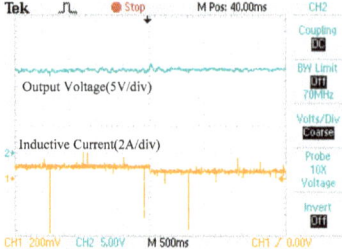

Figure 15. Experimental step-load waveforms of the output voltage and inductive current under the PID controller.

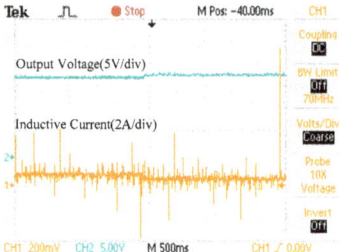

Figure 16. Experimental step-load waveforms of the output voltage and inductive current under the TSM controller.

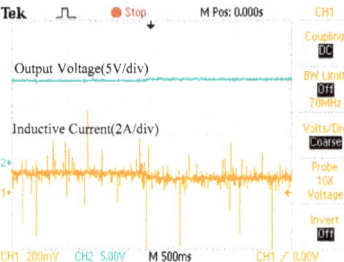

Figure 17. Experimental step-load waveforms of the output voltage and inductive current under the TSM + DOB controller.

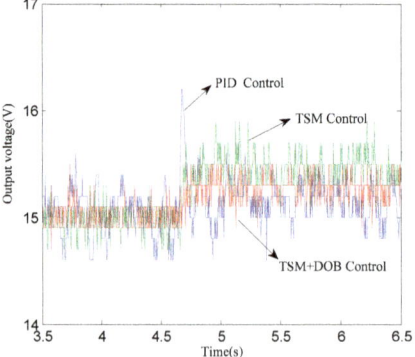

Figure 18. Experimental step-load comparisons of the output voltage under PID, TSM, and TSM + DOB control modes.

Table 3 gives the comparisons of overshoot and settling time under PID, TSM, and TSM + DOB control modes when the step-load changes. It can be seen that the overshoot under the TSM + DOB controller is smaller than that under the other two control modes. The settling time under the TSM + DOB controller is also shorter than that under the other two control modes, with the related rates of 218.0% and 14.8% when comparing with the PID and TSM control mode, respectively.

Table 3. Comparisons under PID, TSM, and TSM + DOB when the step-load changes.

Controller	Overshoot (V)	Settling Time (ms)
PID	1.2	294.5
TSM	0.6	106.3
TSM + DOB	0.4	92.6

Experimental DC coupled step-input-voltage waveforms of the output voltage and inductive current under three control schemes (PID, TSM, TSM + DOB) when the input voltage steps from 30 V to 35 V are shown in Figures 19–21, and the experimental step-input-voltage comparisons of the output voltage are given in Figure 22.

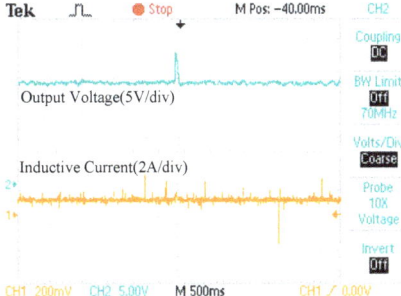

Figure 19. Experimental step-input-voltage waveforms of the output voltage and inductive current under the PID controller.

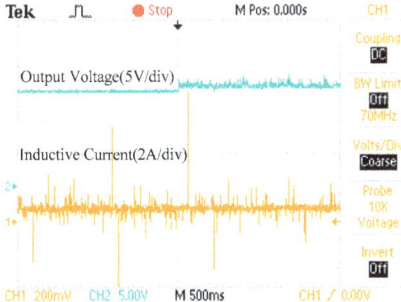

Figure 20. Experimental step-input-voltage waveforms of the output voltage and inductive current under the TSM controller.

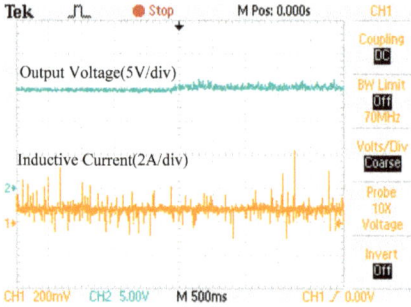

Figure 21. Experimental step-input-voltage waveforms of the output voltage and inductive current under the TSM + DOB controller.

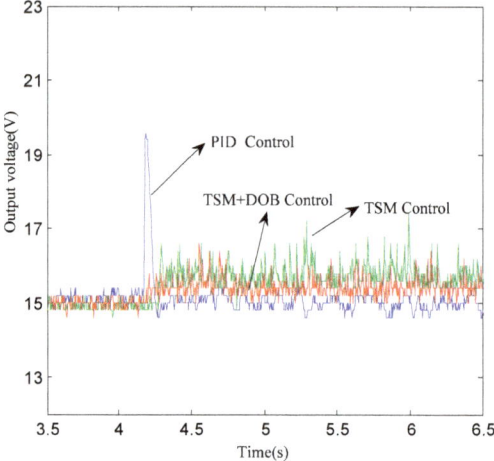

Figure 22. Experimental step-input-voltage comparisons of the output voltage under PID, TSM, and TSM + DOB control modes.

Table 4 gives the comparisons of overshoot and settling time under PID, TSM, and TSM + DOB control modes when the step-input-voltage changes. It can be seen that the overshoot under the TSM + DOB controller is smaller than that under the other two control modes. Moreover, the settling time is also shorter than that under the other two control modes, with the related rates of 59.1% and 20.6% when comparing with the PID and TSM control modes, respectively.

Table 4. Comparisons under PID, TSM, and TSM + DOB when the step-input-voltage changes.

Controller	Overshoot (V)	Settling Time (ms)
PID	4.6	112.6
TSM	1.2	85.4
TSM + DOB	1.0	70.8

From the above experimental results, we can see that in the absence of external disturbances, the terminal sliding mode controller combined with the disturbance observer can quickly reach the desired value, with the overshoot and convergence time being significantly less than those achieved under the other two control schemes. Furthermore, when some disturbances exist, such as the step-load and input-voltage changes, the output voltage of the Buck converter under the compound controller provides the best robustness property. This was confirmed by comparing the overshoot and settling time of the compound controller with those of the other two controllers. The compound controller was found to improve the closed-loop system of Buck converters in two aspects. One is the convergent speed, which implies that the output voltage will converge to the desired voltage rapidly at the start-up. The other is the disturbance rejection property, which provides the control system with strong robustness.

6. Conclusions

In this paper, a new compound control scheme based on terminal sliding mode and nonlinear disturbance observer was proposed for a Buck converter. In comparison to the traditional PID and terminal sliding mode state feedback control modes, this method can provide faster convergence performance and higher voltage output quality for the closed-loop system of a Buck converter. Simulation and experimentation results showed that under the compound control scheme, the system performance of the Buck converter was improved effectively and its robustness was further improved. It can be seen from the experimentations that there always exists a larger steady-state error for the proposed three kinds of controllers. This may be because the power of the Buck converter is lower while the external disturbance is comparatively large. To this end, our future work will focus on DC-DC Buck converters with more power to further test the robustness performances of TSM + DOB algorithms, considering frequency response, dynamic response, immunity to noise, and large-signal stability.

Author Contributions: Conceptualization, L.M.; Methodology and Software, S.D.; Investigation and Writing—Review and Editing, Y.S.; Supervision, D.Z.

Acknowledgments: This work was supported in part by the National Nature Science Foundation of China under Grant 31571571, in part by the Priority Academic Program Development of Jiangsu Higher Education Institutions (PAPD), and in part by the Jiangsu Province Natural Science Fund Project under Grant BK20170536.

Conflicts of Interest: The authors declare no conflict of interest.

References

1. Chen, J.J.; Hsu, J.H.; Hwang, Y.S.; Yu, C.C. A DC–DC buck converter with load-regulation improvement using dual-path-feedback techniques. *Analog. Integr. Circuits Signal Proc.* **2014**, *79*, 149–159. [CrossRef]
2. Ding, S.H.; Zheng, W.X.; Sun, J.; Wang, J. Second-order sliding mode controller design and its implementation for buck converters. *IEEE Trans. Ind. Inf.* **2018**, *14*, 1990–2000. [CrossRef]
3. Shabestari, P.M.; Gharehpetian, G.B.; Riahy, G.H.; Mortazavian, S. Voltage controllers for DC-DC boost converters in discontinuous current mode. In Proceedings of the International Energy and Sustainability conference (IESC), New York, NY, USA, 12–13 November 2015; pp. 1–7.
4. Kancherla, S.; Tripathi, R.K. Nonlinear average current mode control for a DC-DC Buck converter. In Proceedings of the IEEE International Conference on Sustainable Energy Technologies, Singapore, 24–27 November 2008; pp. 831–836.
5. Tan, S.C.; Lai, Y.M.; Cheung, M.K.H.; Tse, C.K. On the practical design of a sliding mode voltage controlled buck converter. *IEEE Trans. Power Electron.* **2005**, *20*, 425–437. [CrossRef]
6. Zhang, Y.; Zhang, B.; Chen, B.; Hu, Z. A Novel Control Law of Boost DC-DC Converter Based on Bilinear Theory. *Trans. China Electrotech. Soc.* **2006**, *21*, 109–114.
7. Tan, S.C.; Lai, Y.M.; Chi, K.T. General Design Issues of Sliding-Mode Controllers in DC-DC Converters. *IEEE Trans. Ind. Electron.* **2008**, *55*, 1160–1174.
8. Ni, Y.; Xu, J.P. Design of a novel discrete global sliding mode controlled Buck converter. *Electr. Mach. Control* **2009**, *23*, 112–116.
9. Komurcugil, H. Non-singular terminal sliding-mode control of DC-DC buck converters. *Control Eng. Pract.* **2013**, *21*, 321–332. [CrossRef]
10. Delgado, F.; Magana, M.E. A fuzzy logic controller design and simulation for a sawmill bucking system. *IEEE Trans. Ind. Electron.* **2005**, *52*, 628–634. [CrossRef]

11. Alrabadi, A.N.; Barghash, M.A.; Abuzeid, O.M. Intelligent Regulation Using Genetic Algorithm-Based Tuning for the Fuzzy Control of the Power Electronic Switching-Mode Buck Converter. Available online: https://pdfs.semanticscholar.org/ea48/74ac9caac6bfccaa70c3b767f54d6eb52a41.pdf (accessed on 12 February 2018).
12. Saravanan, A.G.; Rajaram, M. Fuzzy controller for dynamic performance improvement of a half-bridge isolated dc–dc converter. *Neurocomputing* **2014**, *140*, 283–290. [CrossRef]
13. Piazza, M.C.D.; Pucci, M.; Ragusa, A.; Vitale, G. Analytical versus neural real-time simulation of a photovoltaic generator based on a DC-DC converter. *IEEE Trans. Ind. Appl.* **2010**, *46*, 2501–2510. [CrossRef]
14. Büngöl, O.; Pacaci, S. A Virtual Laboratory for Neural Network Controlled DC Motors Based on a DC-DC Buck Converter. *Int. J. Eng. Educ.* **2012**, *28*, 713–723.
15. Salimi, M.; Soltani, J.; Markadeh, G.A.; Abjadi, N.R. Adaptive nonlinear control of the DC-DC buck converters operating in CCM and DCM. *Eur. Trans. Electr. Power* **2012**, *23*, 1536–1547. [CrossRef]
16. Babazadeh, A.; Maksimovic, D. Hybrid Digital Adaptive Control for Fast Transient Response in Synchronous Buck DC–DC Converters. *IEEE Trans. Power Electron.* **2009**, *24*, 2625–2638. [CrossRef]
17. Vatankhah, B.; Farrokhi, M. Offset-free adaptive nonlinear model predictive control with disturbance observer for dc-dc buck converters. *Turk. J. Electr. Eng. Comput. Sci.* **2017**, *25*, 2195–2206. [CrossRef]
18. Ding, S.H.; Wang, J.D.; Zheng, W.X. Second-order sliding mode control for nonlinear uncertain systems bounded by positive functions. *IEEE Trans. Ind. Electron.* **2015**, *62*, 5899–5909. [CrossRef]
19. Ding, S.H.; Liu, L.; Zheng, W.X. Sliding mode direct yaw-moment control design for in-wheel electric vehicles. *IEEE Trans. Ind. Electron.* **2017**, *64*, 6752–6762. [CrossRef]
20. Xie, X.Z. Observer-based nonsingular terminal sliding mode controller design. In Proceedings of the 24th Chinese Control and Decision Conference (CCDC), Taiyuan, China, 23–25 May 2012; pp. 454–458.
21. Qi, W.H.; Zong, G.D.; Karim, H.R. Observer-based adaptive SMC for nonlinear uncertain singular semi-Markov jump systems with applications to DC motor. *IEEE Trans. Circuit. Syst. I Regul. Pap.* **2018**, *65*, 2951–2960. [CrossRef]
22. Ni, Y.; Xu, J.P.; Wang, J.P.; Yu, H.K. Design of global sliding mode control Buck converter with hysteresis modulation. *Proc. Chin. Soc. Electr. Eng.* **2010**, *30*, 1–6.
23. Naik, B.B.; Mehta, A.J. Sliding mode controller with modified sliding function for DC-DC Buck Converter. *ISA Trans.* **2017**, *70*, 1–5. [CrossRef] [PubMed]
24. Komurcugil, H. Adaptive terminal sliding-mode control strategy for DC-DC buck converters. *ISA Trans.* **2012**, *51*, 673–681. [CrossRef] [PubMed]
25. Wang, J.X.; Li, S.H.; Yang, J.; Wu, B.; Li, Q. Finite-time disturbance observer based non-singular terminal sliding-mode control for pulse width modulation based dc–dc buck converters with mismatched load disturbances. *IET Power Electron.* **2016**, *9*, 1995–2002. [CrossRef]
26. Zhao, Z.; Yang, J.; Li, S.; Yu, X.; Wang, Z. Continuous output feedback TSM control for uncertain systems with a DC-AC inverter example. *IEEE Trans. Circuit. Syst. II Express Br.* **2018**, *65*, 71–75. [CrossRef]
27. Zhang, C.L.; Wang, J.X.; Li, S.H.; Wu, B.; Qian, C.J. Robust control for PWM-based DC–DC buck power converters with uncertainty via sampled-data output feedback. *IEEE Trans. Power Electron.* **2015**, *30*, 504–515. [CrossRef]
28. Yang, J.; Ding, Z. Global output regulation for a class of lower triangular nonlinear systems: A feedback domination approach. *Automatica* **2017**, *76*, 65–69. [CrossRef]
29. Ding, S.H.; Li, S.H. Second-order sliding mode controller design subject to mismatched term. *Automatica* **2017**, *77*, 388–392. [CrossRef]
30. Du, H.B.; Yu, X.H.; Chen, M.Z.Q.; Li, S.H. Chattering-free discrete-time sliding mode control. *Automatica* **2016**, *68*, 87–91. [CrossRef]
31. Wang, J.; Liang, K.; Huang, X.; Wang, Z.; Shen, H. Dissipative fault-tolerant control for nonlinear singular perturbed systems with markov jumping parameters based on slow state feedback. *Appl. Math. Comput.* **2018**, *328*, 247–262. [CrossRef]
32. Shen, H.; Li, F.; Xu, S.; Sreeram, V. Slow state variables feedback stabilization for semi-markov jump systems with singular perturbations. *IEEE Trans. Autom. Control* **2017**, *63*, 2709–2714. [CrossRef]
33. Cheng, J.; Chang, X.H.; Ju, H.P.; Li, H.; Wang, H. Fuzzy-model-based H_∞ control for discrete-time switched systems with quantized feedback and unreliable links. *Inf. Sci.* **2018**, *436–437*, 181–196. [CrossRef]

34. Bhat, S.P. Finite-time stability of continuous autonomous systems. *SIAM J. Control Optim.* **2000**, *38*, 751–766. [CrossRef]
35. Cheng, J.; Ju, H.P.; Karimi, H.R.; Hao, S. A flexible terminal approach to sampled-data exponentially synchronization of Markovian neural networks with time-varying delayed signals. *IEEE Trans. Cybern.* **2018**, *48*, 2232–2244. [PubMed]
36. Ge, C.; Wang, B.; Wei, X.; Liu, Y. Exponential synchronization of a class of neural networks with sampled-data control. *Appl. Math. Comput.* **2017**, *315*, 150–161. [CrossRef]
37. Zhang, C.L.; Yan, Y.D.; Narayan, A.; Yu, H.Y. Practically oriented finite-time control design and implementation: Application to series elastic actuator. *IEEE Trans. Ind. Electron.* **2018**, *65*, 4166–4176. [CrossRef]
38. Li, S.H.; Yang, J.; Chen, W.H.; Chen, X. *Disturbance Observer Based Control: Methods and Applications*; CRC Press: Boca Raton, FL, USA, 2014.

© 2018 by the authors. Licensee MDPI, Basel, Switzerland. This article is an open access article distributed under the terms and conditions of the Creative Commons Attribution (CC BY) license (http://creativecommons.org/licenses/by/4.0/).

Article

Disturbance-Observer-Based Model Predictive Control for Battery Energy Storage System Modular Multilevel Converters

Yantao Liao [1], Jun You [1,*], Jun Yang [2], Zuo Wang [2] and Long Jin [1]

1. School of Electrical Engineering, Southeast University, Nanjing 210096, China; liaoyt@seu.edu.cn (Y.L.); jinlong@seu.edu.cn (L.J.)
2. Key Laboratory of Measurement and Control of Complex Systems of Engineering, Ministry of Education, School of Automation, Southeast University, Nanjing 210096, China; j.yang84@seu.edu.cn (J.Y.); z.wang@seu.edu.cn (Z.W.)
* Correspondence: youjun@seu.edu.cn; Tel.: +86-136-0519-1945

Received: 31 July 2018; Accepted: 28 August 2018; Published: 30 August 2018

Abstract: Although the traditional model predictive control (MPC) can theoretically provide AC current and circulating current control for modular multilevel converters (MMCs) in battery energy storage grid-connected systems, it suffers from stability problems due to the power quality of the power grid and model parameter mismatches. A two discrete-time disturbance observers (DOBs)-based MPC strategy is investigated in this paper to solve this problem. The first DOB is used to improve the AC current quality and the second enhances the stability of the circulating current control. The distortion and fluctuation of grid voltage and inductance parameter variation are considered as lump disturbances in the discrete modeling of a MMC. Based on the proposed method, the output prediction is compensated by disturbance estimation to correct the AC current and circulating current errors, which eventually achieve the expected tracking performance. Moreover, the DOBs have a quite low computational cost with minimum order and optimal performance properties. Since the designed DOBs work in parallel with the MPC, the control effect is improved greatly under harmonics, 3-phase unbalance, voltage sag, inductance parameter mismatches and power reversal conditions. Simulation results confirm the validity of the proposed scheme.

Keywords: battery energy storage system; modular multilevel converter; model predictive control; disturbance observer; AC current control; circulating current control

1. Introduction

The increasing popularity of clean energy has prompted new attention to the stability of the power system because of its intermittent character. The microgrid based on battery energy storage can restrain the negative influences of renewable power supplies on the power grid. Therefore, it is significant to research the battery energy storage technology [1]. As the interface to the power grid, the grid-connected converter for battery energy storage system determines the effect of bidirectional power exchange, AC/DC current control and discharge/recharge [2–4]. However, there are various non-determinacies and disturbances that widely exist in the system, including harmonics, 3-phase unbalance, voltage sag from the power grid, and model parameter mismatches, etc. They usually have negative effects on the stability and performance of control systems [5–8].

Different types of converter topologies, containing the traditional two-level converter, multi-module converter, multilevel converter, and the newly recently advanced modular multilevel converter (MMC), have been presented and studied for battery storage converter applications [9–13]. Amid all the topology structures, the MMC is a recommended topology due to its merits such as high quality of the output voltage, flexible scaling of sub-modules (SMs), bidirectional power flow, and independent regulation of the active and reactive power [14,15]. These inherently salient features make the MMC ideally suited for battery energy storage systems.

Actually, the MMC system has the characteristics of large range of operational points and highly nonlinearity, which leads to the urgent need to design effective controllers and develop proper modulation techniques. Hence, many researchers have focused on the control issues of MMCs in recent years. In [16] a novel control method for a d-q frame based model of the MMC is proposed. Based on the reference values of six independent dynamical state variables, the MMC switching functions under steady state were obtained and a direct Lyapunov method was used to improve the dynamic components of those functions. A multi-loop control strategy based on a six-order dynamic model of the MMC was presented in [5] to ensure stable operation under both load and MMC parameters variations. In [17], differential flatness theory was employed to control a flat outputs-based dynamic model of the MMC innovatively, which greatly improves the MMC power sharing ability and enhances robustness. In addition, a novel modulation function-based controller which achieves two separate modulation functions to generate the switching signals of upper and lower SMs was proposed in [8,18]. This method not only maintains stable operation under varying parameter conditions, but also is easily applied.

However, the existing methods which need to modulate the switching signals are not straightforward and fast when compared with model predictive control (MPC) [19]. As a nonlinear control method, MPC is of great interest for MMC applications because of its simple modulation approach, flexible control goals and easy inclusion of nonlinearities. The best switch states are obtained by cost functions in the MPC method. Despite the excellent prospects and beneficial results obtained in the application of MPC to MMC, it is relatively slow due to its feedback regulation to asymptotically suppress the uncertainties and disturbances [20]. Thus, the control performance of the MPC is severely hindered in the presence of model uncertainties and external disturbances. Note that integral action is employed in MPC to eliminate the influences of uncertainties and disturbances [21], but the integrator would sacrifice original control effects because the integral term has coupled interactions with other properties, such as dynamic responses and stability.

The disturbance observer (DOB), an effective approach to compensate the effects of model uncertainties and external disturbances, is widely applied in electric power systems [22–25]. The main characteristic of DOB is that the robustness is achieved without sacrificing any of the original control performance [26]. In addition, this method does not need to establish an accurate mathematical model for the disturbance signals, and its structure is relatively simple, thus, the complicated calculations are avoided in the prediction of disturbance signals, which is beneficial to meet the requirements of real time applicability. In view of this, the combination of MPC with DOB has been extensively studied to improve the anti-disturbance capabilities of various systems, e.g., induction machine drives [24], permanent magnet synchronous motors [27], neutral-point-clamped multilevel converters [28], and three-phase inverters [6,7,29]. However, there are still many gaps in the MMC applications in battery energy storage systems.

As a grid-connected converter of a battery energy storage system, the grid voltage directly affects the control performance. The output current can be distorted by the power quality problems of the grid voltage such as harmonics, 3-phase unbalance and voltage sag. Then, the distorted current that is injected into the power grid would further damage the power quality. On the other hand, AC current control and circulating current suppression for MMC can be influenced by the varying parameters of resistances and inductances which are easily affected by temperature, frequency and electrical life. Hence, the control system needs effective real-time compensation for these uncertainties and disturbances [6,17].

In this paper, a MPC strategy with two linear discrete-time DOBs is proposed for an eleven-level MMC-based battery energy storage grid-connected converter. The MPC strategy divides the cost functions into three types, according to the three control objectives of AC current, circulating current and capacitor voltage-balancing [30]. Two DOBs are designed on the basis of the first two control purposes. The first DOB increases the anti-disturbance capacity of AC current control against the fluctuations of the grid voltage and inductance values. The second enhances the stability of circulating current control under the varying inductance value condition. The capacitor voltage-balancing control is the same as described in [30]. The main contributions of this paper include the following:

(1) A MPC method based on prediction accuracy improvement via two DOBs is designed for a grid-connected MMCs of battery energy storage systems, which has a simple structure and quite low cost of computation with the minimum order.
(2) The accurate estimation and feedforward compensation for disturbances are achieved without sacrificing the original control performance.
(3) The disturbance items which act on the cost functions during each sampling period assure the cost functions always maintain optimal performance.

The studies are implemented based on time-domain simulations in the MATLAB/Simulink environment for five different operating conditions. Through the comparative results, it is validated that the proposed method works reliably even under power grid uncertainty and model parameter mismatch conditions.

2. MPC Strategy of the MMC

Figure 1 presents the circuit diagram of a 3-phase MMC, which is composed of two arms per phase. Each arm comprises $N/2$ series-connected half-bridge SMs, where N is the amount of SM per each phase, and a series inductor L. The SM consists of two IGBTs and one capacitor. The two switches (T_1 and T_2) per SM are controlled with complementary gating signals, resulting in two switch states which can input or bypass the capacitor. Thus, the voltage of the SM depends on the active switching states. The output voltage is the same as the capacitor when the SM is ON. In contrast, when it is OFF, the SM voltage turns to zero.

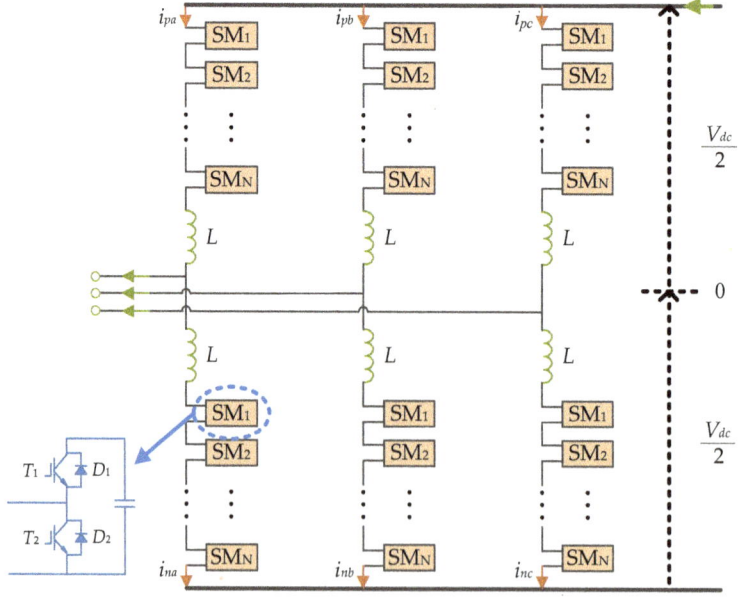

Figure 1. Circuit diagram of three-phase MMC.

Figure 2 shows the equivalent model which is a simplification of 3-phase MMC presented in Figure 1. It describes one phase of the 3-phase MMC connected to the power grid across a filter inductor. The SMs in each arm are replaced by an AC power source and the DC-bus is equivalent to two DC power sources in series with the ground point.

Here, the upper-arm and lower-arm current are described by i_{pk} and i_{nk} ($k = a, b, c$), respectively, where i_{diffk} is the inner unbalanced current; V_k is the grid voltage; and e_{pk} and e_{nk} are the upper-arm and lower-arm voltage, respectively. The current that flows through the arms comprises the DC component and AC component of fundamental frequency in the ideal case, but the capacitor voltages are time-varying due to the AC current flow across the capacitors. As a result, there is a voltage difference between the arm voltage and DC voltage, which causes a circulating current in the converter [31].

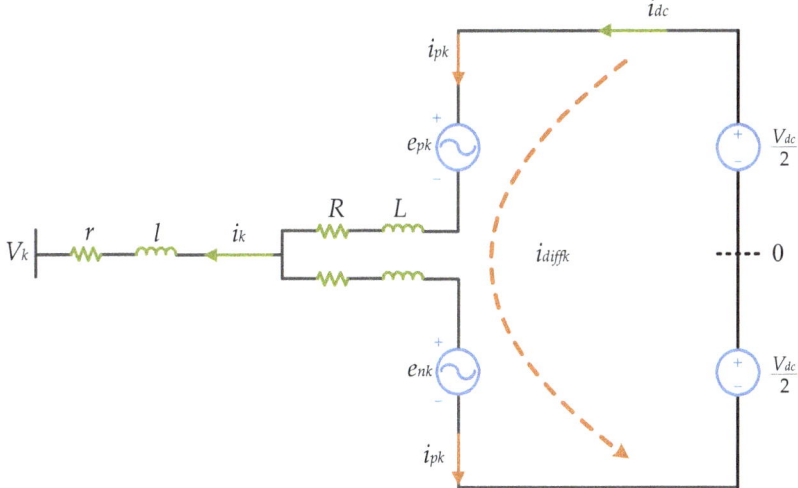

Figure 2. Model of single-phase of the three-phase MMC.

The MPC method has three control objectives, thus, three cost functions are calculated according to them. They include: AC current control; circulating current suppression; and SM capacitor voltage balancing. In this paper, AC current control and circulating current suppression are considered and redesigned based on the disturbances and uncertainties of grid voltage and inductances.

2.1. MPC Strategy for AC Current

From Figure 2, the voltage equation of MMC can be expressed as follows:

$$L' \frac{di_k}{dt} = e_k - V_k - R' i_k \tag{1}$$

where $L' = l + L/2$; $R' = r + R/2$; and e_k is the converter output voltage, defined as:

$$e_k = \frac{e_{nk} - e_{pk}}{2} \tag{2}$$

where $e_{pk} = V_{dc}/2 - e_k$; and $e_{nk} = V_{dc}/2 + e_k$.

The AC current are deduced in Equation (3) using the Euler forward equation:

$$i_k(t + T_s) = \frac{T_s}{L'}[e_k(t + T_s) - V_k(t)] + \left(1 - \frac{T_s R'}{L'}\right) i_k(t) \tag{3}$$

where T_s is the sampling period, which is considered extremely short.

The cost function can be designed by:

$$j_k = |i_k^*(t + T_s) - i_k(t + T_s)| \tag{4}$$

where $i_k^*(t + T_s)$ is the AC current reference.

2.2. MPC Strategy for Circulating Current

The voltage and inner unbalanced current Equations are described as follows:

$$\frac{V_{dc}}{2} - \frac{e_{pk} + e_{nk}}{2} = Ri_{diffk} + L\frac{di_{diffk}}{dt} \tag{5}$$

$$i_{diffk} = \frac{i_{pk} + i_{nk}}{2} \tag{6}$$

The inner unbalanced current is derived from the Euler forward equation:

$$i_{diffk}(t + T_s) = \frac{T_s}{2L}\left\{V_{dc}(t) - \left[e_{pk}(t + T_s) + e_{nk}(t + T_s)\right]\right\} + \left(1 - \frac{T_s R}{L}\right)i_{diffk}(t) \tag{7}$$

Equation (2) shows that e_k depends on the difference between upper-arm and lower-arm voltage. Thus, the same voltage V_{diffk} added to e_{pk} and e_{nk} has no effect on the AC current. Hence, Equation (7) is rewritten as follows:

$$i_{diffk}(t + T_s) = \frac{T_s}{2L}\left\{V_{dc}(t) - \left[(e_{pk}(t + T_s) + V_{diffk}) + (e_{nk}(t + T_s) + V_{diffk})\right]\right\} + \left(1 - \frac{T_s R}{L}\right)i_{diffk}(t) \tag{8}$$

Two voltage levels are allocated to V_{diffk} as in Equation (9). It can be expanded according to the characteristics of the circulating current:

$$V_{diffk} = \left[\frac{V_{dc}}{N}\right][-1, 0, 1] \tag{9}$$

Assuming there is no loss in MMC and the active power is controlled without ripples. The AC active power and DC active power of MMC are expressed as Equation (10) and the DC current reference is obtained as Equation (11) [32]. Thus, the cost function is designed as:

$$P_{ac} = P_{dc} = V_{dc} * i_{dc} \tag{10}$$

$$i_{dc}^* = \frac{P_{ac}}{V_{dc}} \tag{11}$$

$$j_{diffk} = \left|\frac{i_{dc}^*(t + T_s)}{3} - i_{diffk}(t + T_s)\right| \tag{12}$$

2.3. MPC Strategy for Capacitor Voltage Balancing

The SM capacitor voltage is calculated by:

$$\text{SM Turn on} : V_{cki}(t + T_s) = V_{cki}(t) + i_m(t)T_s/C \tag{13}$$

$$\text{SM Turn off} : V_{cki}(t + T_s) = V_{cki}(t) \tag{14}$$

where $i_m(t) = i_{pk}$ or i_{nk}, with $I = 1, 2, 3, \ldots, N$.

The cost function can be designed by:

$$j_{Vck} = \sum_{i=1}^{N} \left| \frac{V_{dc}}{N} - V_{cki}(t+T_s) \right| \quad (15)$$

3. MPC Strategy of the MMC with Disturbance Observer

The MPC controller could be affected by the disturbances caused by grid voltage and inductances. In this section, two DOB-based control structures are described to track the disturbances. The first DOB is designed to improve the waveform quality of AC current and the second is applied to suppress the circulating current. The DOBs to be designed are easy to assign the exponential stability of the estimation error dynamics and the order is minimal [33].

3.1. MPC Strategy for AC Current with DOB 1

Considering Equation (3) by neglecting the influence of resistances, letting:

$$x_{k,n+1}^{(1)} = i_k(t+T_s) \quad (16)$$

$$x_{kn}^{(1)} = i_k(t) \quad (17)$$

$$u_{kn}^{(1)} = e_k(t+T_s) - V_k(t) \quad (18)$$

$$d_{kn}^{(1)} = \frac{1}{L'+d_{L'}} \{e_k(t+T_s) - [V_k(t)+d_{vk}]\} - \frac{1}{L'}[e_k(t+T_s) - V_k(t)] \quad (19)$$

where $d_{kn}^{(1)}$ is a unknown value of disturbances in Equation (3) that needs to be observed; $d_{L'}$ is the disturbances of system inductances; and d_{vk} is the disturbances of grid voltage.

Consequently, the model of Equation (3) can be expressed in a compact form as follows:

$$\begin{cases} x_{k,n+1}^{(1)} = \Phi_1 x_{kn}^{(1)} + \Gamma_1 u_{kn}^{(1)} + G_1 d_{kn}^{(1)}, x_{k0}^{(1)} = x_k^{(1)}(0) \\ y_{kn}^{(1)} = C_1 x_{kn}^{(1)} \end{cases} \quad (20)$$

where $\Phi_1 = 1; \Gamma_1 = T_s/L'; G_1 = T_s;$ and $C_1 = 1$.

Thus, the first discrete-time disturbance observer is applied to estimate the disturbances of system (20), given by:

$$\begin{cases} \hat{d}_{kn}^{(1)} = K_1 x_{kn}^{(1)} - z_{kn}^{(1)} \\ z_{k,n+1}^{(1)} = z_{kn}^{(1)} + K_1 \left[(\Phi_1 - 1) x_{kn}^{(1)} + \Gamma_1 u_{kn}^{(1)} + G_1 \hat{d}_{kn}^{(1)} \right] \end{cases} \quad (21)$$

where $\hat{d}_{kn}^{(1)}$ is the estimated value of $d_{kn}^{(1)}$; $z_{kn}^{(1)}$ is the state variable of DOB 1; and K_1 is the coefficient of the pole assignment of DOB 1 to be designed.

Then, generally, the state estimation error, $e_{n+1} \triangleq d_n - \hat{d}_n$, has the dynamics:

$$e_{n+1} = (1-KG)e_n + \Delta d_{n+1} \quad (22)$$

where $\Delta d_{n+1} = d_{n+1} - d_n$.

Letting $\lambda = 1 - KG$, the estimation error is confined within a bound such that:

$$|e_\infty| \le \frac{T_s \mu}{1-|\lambda|} \quad (23)$$

where $|\lambda| \le 1$; and μ is a small positive value.

Hence, the coefficient of the pole assignment of DOB 1, K_1 can be given by:

$$K_1 = \frac{(1-\lambda_1)}{G_1}, \ |\lambda_1| \leq 1 \tag{24}$$

3.2. MPC Strategy for Circulating Current with DOB 2

Considering Equation (8) by neglecting the influence of resistances, letting:

$$x_{k,n+1}^{(2)} = i_{diffk}(t+T_s) \tag{25}$$

$$x_{kn}^{(2)} = i_{diffk}(t) \tag{26}$$

$$u_{kn}^{(2)} = V_{dc}(t) - \left[(e_{pk}(t+T_s) + V_{diffk}) + (e_{nk}(t+T_s) + V_{diffk})\right] \tag{27}$$

$$d_{kn}^{(2)} = \left(\frac{1}{L+d_L} - \frac{1}{L}\right)\left\{V_{dc}(t) - \left[(e_{pk}(t+T_s) + V_{diffk}) + (e_{nk}(t+T_s) + V_{diffk})\right]\right\} \tag{28}$$

where $d_{kn}^{(2)}$ is a unknown value of disturbances in Equation (8) that needs to be observed; and d_L is the disturbances of arm inductances.

Similarly, the model (8) can be rewritten as follows:

$$\begin{cases} x_{k,n+1}^{(2)} = \Phi_2 x_{kn}^{(2)} + \Gamma_2 u_{kn}^{(2)} + G_2 d_{kn}^{(2)}, x_{k0}^{(2)} = x_k^{(2)}(0) \\ y_{kn}^{(2)} = C_2 x_{kn}^{(2)} \end{cases} \tag{29}$$

where $\Phi_2 = 1$; $\Gamma_2 = T_s/2L$; $G_2 = T_s/2$; and $C_2 = 1$.

Consequently, the second discrete-time disturbance observer is applied to system (29), given by:

$$\begin{cases} \hat{d}_{kn}^{(2)} = K_2 x_{kn}^{(2)} - z_{kn}^{(2)} \\ z_{k,n+1}^{(2)} = z_{kn}^{(2)} + K_2\left[(\Phi_2-1)x_{kn}^{(2)} + \Gamma_2 u_{kn}^{(2)} + G_2 \hat{d}_{kn}^{(2)}\right] \end{cases} \tag{30}$$

$$K_2 = \frac{(1-\lambda_2)}{G_2}, \ |\lambda_2| \leq 1 \tag{31}$$

where $\hat{d}_{kn}^{(2)}$ is the estimated value of $d_{kn}^{(2)}$; $z_{kn}^{(2)}$ is the state variable of DOB 2; and K_2 is the coefficient of the pole assignment of DOB 2.

Finally, the AC current and inner unbalanced current with DOBs can be given by:

$$i_k(t+T_s) = \frac{T_s}{L'}[e_k(t+T_s) - V_k(t)] + \left(1 - \frac{T_s R'}{L'}\right)i_k(t) + \hat{d}_{kn}^{(1)} \tag{32}$$

$$i_{diffk}(t+T_s) = \frac{T_s}{2L}\left\{V_{dc}(t) - \left[(e_{pk}(t+T_s) + V_{diffk}) + (e_{nk}(t+T_s) + V_{diffk})\right]\right\} + (1 - \frac{T_s R}{L})i_{diffk}(t) + \hat{d}_{kn}^{(2)} \tag{33}$$

In summary, $\hat{d}_{kn}^{(1)}$ and $\hat{d}_{kn}^{(2)}$ work as patches of MPC. They are decided by the actual system and equal to zero in the absence of disturbance and uncertainties. Thus, the DOB-based MPC control system would not be worse than the original MPC controller. On the other hand, they are calculated with the cost function simultaneously, which guarantees the cost functions always maintain their optimal performance no matter how the actual system chang.

4. Simulation Results

Simulations were carried out by using MATLAB/Simulink, and Figure 3 shows the structure of simulated system and proposed control method with DOBs. The critical parameters of circuit are reported in Table 1, and the Table 2 lists the control parameters of DOB 1 and DOB 2.

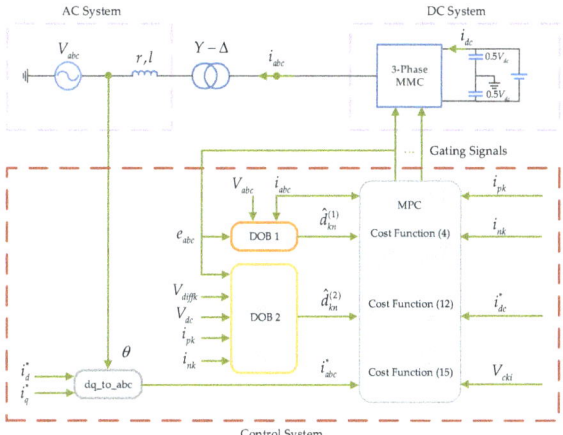

Figure 3. System structure of simulation.

Table 1. Main circuit parameters.

Parameters	Values	Units
Rated Power	1.2	MW
AC System Voltage	9800	V
Line Frequency	50	Hz
AC System Inductance	2	mH
DC Bus Voltage	20	kV
Number of SMs per arm	10	-
SM Capacitance	0.002	F
SM Capacitor Voltage	2000	V
Arm Inductance	0.02	H
Sampling and control period	20	μs

Table 2. Control parameters of DOB 1 and DOB 2.

Parameters	DOB 1	DOB 2
Φ	1	1
Γ	0.0017	0.0005
G	0.00002	0.00001
C	1	1
K	40,000	100,000
λ	0.2	1

The MPC Strategy with DOB has been analyzed for five different test conditions:

- Harmonic: 3-phase grid voltage with 30% of 5th and 7th harmonics.
- 3-phase voltage unbalance: a-phase line-to-ground fault.
- Voltage sag: 3-phase grid voltage with 80% of reduction in the period of 0.01 s–0.03 s.
- Parameter mismatches: actual inductance values change.
- Power reversal: the power flow reverses from 0.05 s to 0.1 s; the 3-phase grid voltage contain harmonics and voltage sag; and the actual inductance values decline during the entire simulation time.

4.1. Simulations under Harmonic Condition

30% of 5th and 7th harmonics are overlaid onto the 3-phase grid voltage. The AC active current reference is 100 A, and the reactive current reference is 0 A. Figure 4a shows the 3-phase grid voltage with 5th and 7th harmonics. Figure 4b shows the 3-phase grid current without DOB. It is obvious from this figure that the grid current has been significantly distorted. The grid current is compensated by the proposed DOB as represented in Figure 4c. Compared with the Figure 4b, it can be seen that the harmonic current is effectively suppressed.

Figure 5 shows the tracing results between the actual disturbances and estimated disturbances. The controller enters the steady state after 1 ms. The estimated value has a certain lag with respect to the actual value, because the estimated output value is filtered by LPF. The cut-off frequency design needs consideration of delay and tracking curve smoothness. It is set to 2000 Hz in this paper. It is clearly shown that two waveforms well coincide with each other, which proves the validity of the proposed DOB.

The comparison of 5th and 7th harmonic current amplitude of a-phase between original controller and proposed controller is given in Figure 6. The further comparisons of grid current harmonics and THD are listed in Table 3. The 5th harmonic current amplitude of a-phase drops from 3.99 A to 0.95 A, and the 7th harmonic current amplitude of a-phase drops from 4.04 A to 1.30 A. It is similar to the other two phases. With the sharp reduction of harmonic current, the THD of 3-phase current also drops from 6.97%, 6.75% and 6.68% to 2.86%, 2.76% and 2.97% respectively.

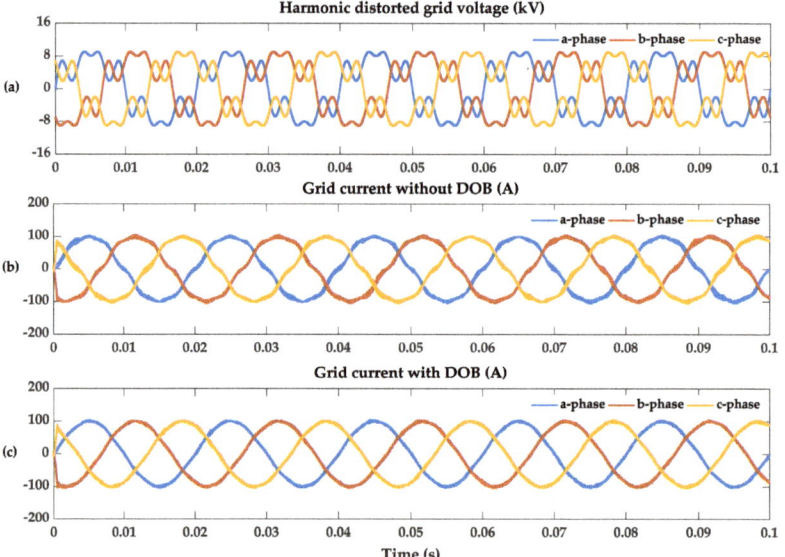

Figure 4. Simulations under harmonic grid voltage condition: (a) Harmonic distorted grid voltage; (b) Grid current without DOB; (c) Grid current with DOB.

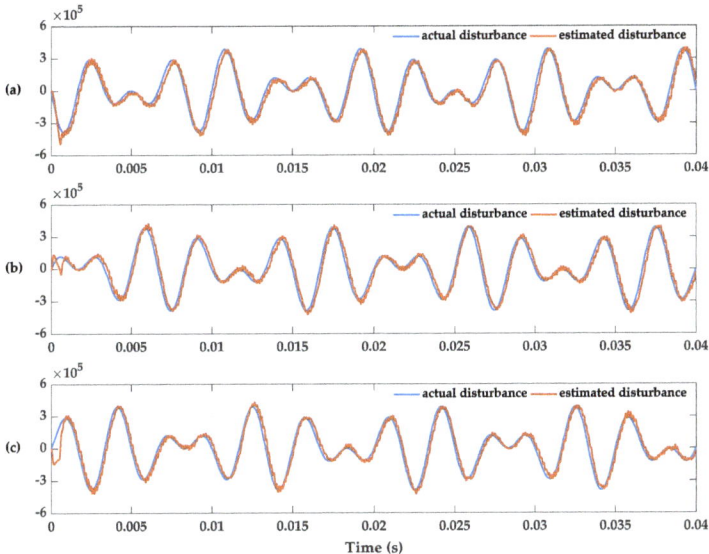

Figure 5. Actual disturbance value $d_{kn}^{(1)}$ and estimated disturbance value $\hat{d}_{kn}^{(1)}$ under harmonic condition: (**a**) A-phase; (**b**) B-phase; (**c**) C-phase.

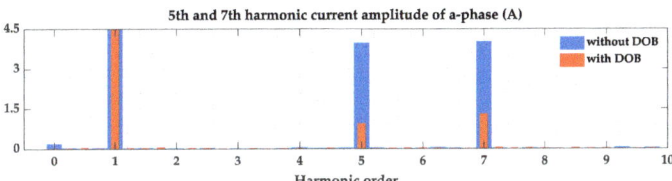

Figure 6. Comparison of 5th and 7th harmonic current amplitude of a-phase between grid current without DOB and with DOB.

Table 3. Comparisons of grid current under harmonic grid voltage condition.

Phase	5th/7th Harmonic (A)		THD (%)	
	Without DOB	With DOB	Without DOB	With DOB
A	3.99/4.04	0.95/1.30	6.97	2.86
B	4.02/4.01	0.90/1.31	6.75	2.76
C	4.03/4.01	0.97/1.31	6.68	2.97

4.2. Simulations Under 3-Phase Voltage Unbalance Condition

A-phase line-to-ground fault is applied in this condition. The AC active current reference is 100 A, and the reactive current reference is 0 A. Figure 7a shows the unbalanced voltage. The a-phase voltage drops to 0 V. Figure 7b shows the gird current without DOB. The a-phase current not only increases by approximately 10 A, but also brings current ripples to the 3-phase current. Figure 7c shows the improved grid current using DOB. It is clearly shown that the a-phase current amplitude is restored to the normal value and the harmonics of 3-phase current are all decreased.

Figure 8 shows the estimation performances of unbalanced voltage disturbances using the proposed DOB. In this case, the tracking results to the actual disturbances are more efficient because the disturbances are gentler than under the harmonic condition. Thus, the inhibiting effect is better.

Table 4 represents the fundamental amplitude and THD of 3-phase current. The fundamental amplitude of a-phase current is up to 109 A, and the other two phases rise 2 A approximately. All of them decline to about 100 A when the proposed DOB is enabled. Similarly, the THD of 3-phase current drops from 4.98%, 3.75% and 3.85% to 2.52%, 2.20% and 2.17%, respectively.

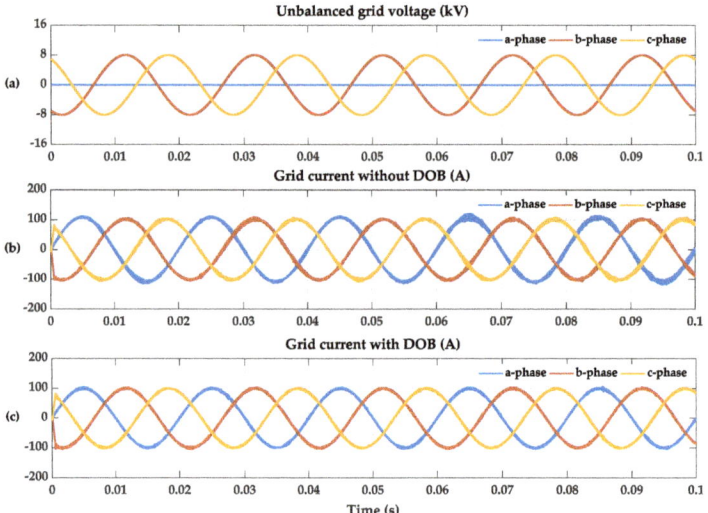

Figure 7. Simulations under unbalanced voltage condition: (**a**) Unbalanced grid voltage; (**b**) Grid current without DOB; (**c**) Grid current with DOB.

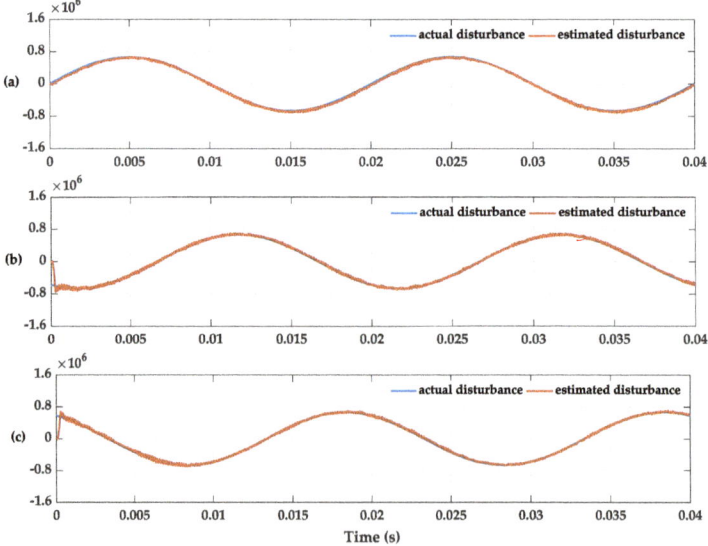

Figure 8. Actual disturbance value $d_{kn}^{(1)}$ and estimated disturbance value $\hat{d}_{kn}^{(1)}$ under unbalanced voltage condition: (**a**) A-phase; (**b**) B-phase; (**c**) C-phase.

Table 4. Comparisons of grid current under unbalanced grid voltage condition.

Phase	Fundamental Amplitude (A)		THD (%)	
	Without DOB	With DOB	Without DOB	With DOB
A	109	99.97	4.98	2.52
B	102.1	100.2	3.75	2.20
C	101.9	99.79	3.85	2.17

4.3. Simulations Under Voltage Sag Condition

3-phase grid voltage sag condition is simulated in this case. The AC active current reference is 100 A, and the reactive current reference is 0 A. The magnitude of grid voltage amplitude drops to 20% at 0.01 s and returns to the normal value at 0.03 s, as shown in Figure 9a. Figure 9b represents the grid current without DOB. It is obvious that the grid current increases and inrush current is generated in the moment of 0.01 s and 0.03 s. Figure 9c illustrates the comparative simulation result for Figure 9b. The grid current amplitude returns to the normal value by applying the proposed DOB compensation scheme as in Figure 9c. But the inrush current still exists.

Figure 10 shows the tracking effect between actual disturbances and estimated disturbances. The estimated disturbance value fluctuates around 0 slightly when the actual disturbances are not yet injected. It proves that the proposed method doesn't sacrifice the original control performances. The fluctuation of a-phase actual disturbance value is smooth at 0.01 s and 0.03 s. But it is not similar to b-phase and c-phase. With the rapid change of actual disturbance value, although the tracing is very fast, the difference between actual and estimated disturbance value is very large in the moment of 0.01 s and 0.03 s. It is the reason for the inrush current.

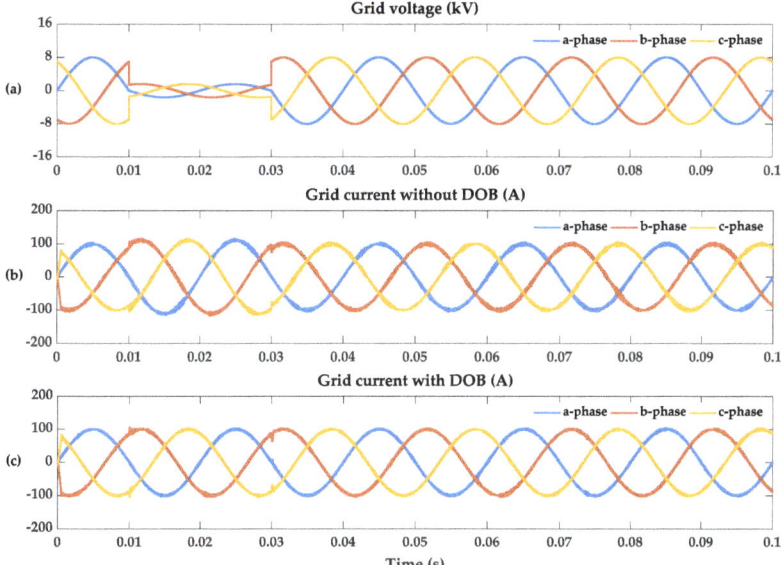

Figure 9. Simulations under voltage sag condition: (a) Grid voltage; (b) Grid current without DOB; (c) Grid current with DOB.

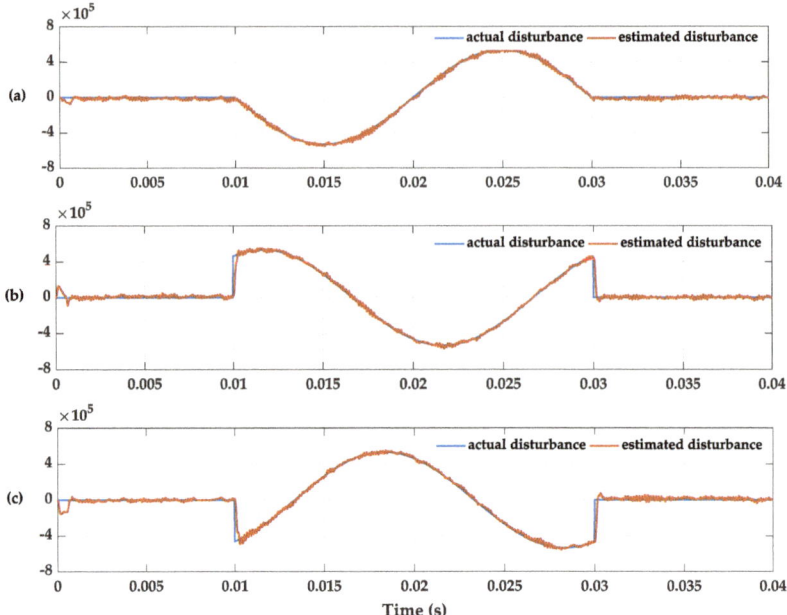

Figure 10. Actual disturbance value $d_{kn}^{(1)}$ and estimated disturbance value $\hat{d}_{kn}^{(1)}$ under voltage sag condition: (**a**) A-phase; (**b**) B-phase; (**c**) C-phase.

4.4. Simulations under Parameter Mismatches Condition

The inductance values are not changeless in a real system, however, the controller parameters are fixed, thus, the controller could be invalid under the inductance value varying condition. At first, a simulation of actual inductance value reduction condition is given. It proves that the proposed MPC controller with DOB can improve the control effect of grid current when the actual values of all the inductances fall by a third. Second, considering the influence of actual inductance value varying on the circulating current is much larger than that of the parameter mismatches, the compensation effect of DOB on circulating current suppression is observed by raising the inductance values by 50 times. The AC active current reference is 100 A, and the reactive current reference is 0 A.

4.4.1. Actual Inductance Value Reduction

The grid current is influenced by the actual inductance value reduction significantly as in Figure 11a. The MPC controller without DOB is affected by the parameter mismatches and there are great ripples in 3-phase current. Figure 11b represents the grid current using the MPC controller with DOB. It is shown that the proposed controller improves the performance of the original under the parameter mismatches condition.

As shown in Figure 12, the 3-phase estimated disturbance value can track the actual disturbance value stably. Furthermore, Table 5 presents the fundamental amplitude and THD of 3-phase current. It turns out that actual inductance value reduction not only brings current ripples, but also lowers the fundamental amplitude. By adding the proposed DOB on the MPC controller, the grid current amplitude returns to normal value and THD reduces significantly.

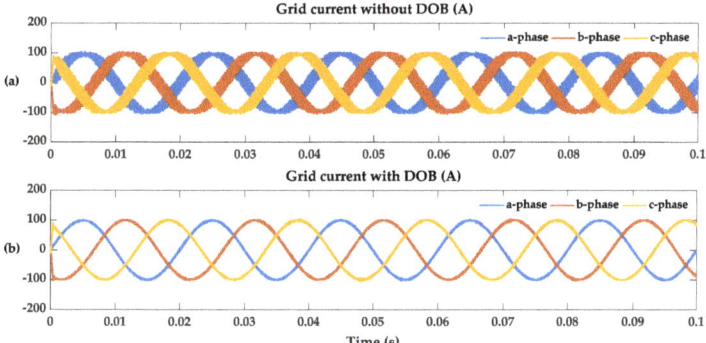

Figure 11. Simulations under actual inductance value reduction condition: (**a**) Grid current without DOB; (**b**) Grid current with DOB.

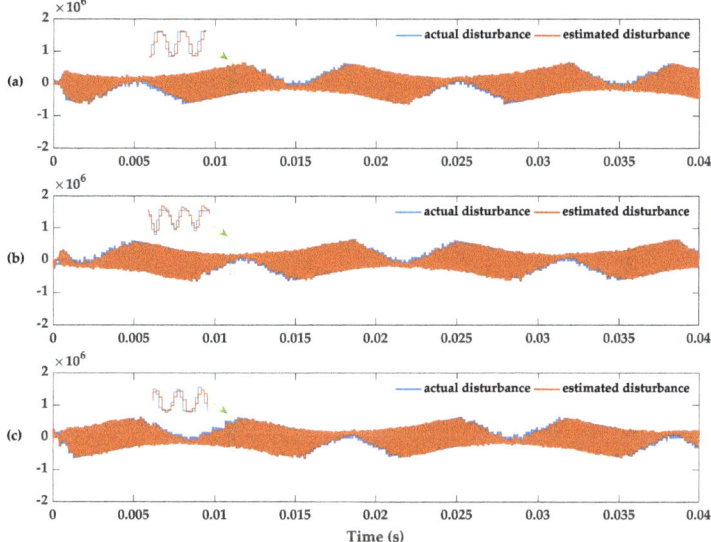

Figure 12. Actual disturbance value $d_{kn}^{(1)}$ and estimated disturbance value $\hat{d}_{kn}^{(1)}$ under actual inductance value reduction condition: (**a**) A-phase; (**b**) B-phase; (**c**) C-phase.

Table 5. Comparisons of grid current under actual inductance value reduction condition.

Phase	Fundamental Amplitude (A)		THD (%)	
	Without DOB	With DOB	Without DOB	With DOB
A	95.66	100	21.56	2.12
B	95.68	99.96	21.44	2.06
C	95.62	100	21.56	2.13

4.4.2. Actual Inductance Values Increase

In order to verify the compensation effect of DOB on circulating current suppression, the actual inductance values are increased by 50 times. Figure 13a shows the tracking of the inner unbalanced current to its reference without DOB. Under the influence of parameter mismatches, the original controller is unable to enter a steady state, and the 3-phase inner unbalanced current which its reference is 20 A keeps fluctuating near 19 A. Shown in Figure 13b is a similar situation where the DC current that fluctuates near 57.5 A can't trace the steady state value of 60 A because of the circulating current. Simulation results of inner unbalanced current and DC current with DOB are presented in Figure 13b,c, respectively. It is shown that the inner unbalanced current and DC current can track their references under control of the proposed method. Figure 14 shows the tracing results between actual disturbance value and estimated disturbance value. It proves that the proposed DOB can observe the change of actual system parameters and make the original MPC controller better by matching with the real system.

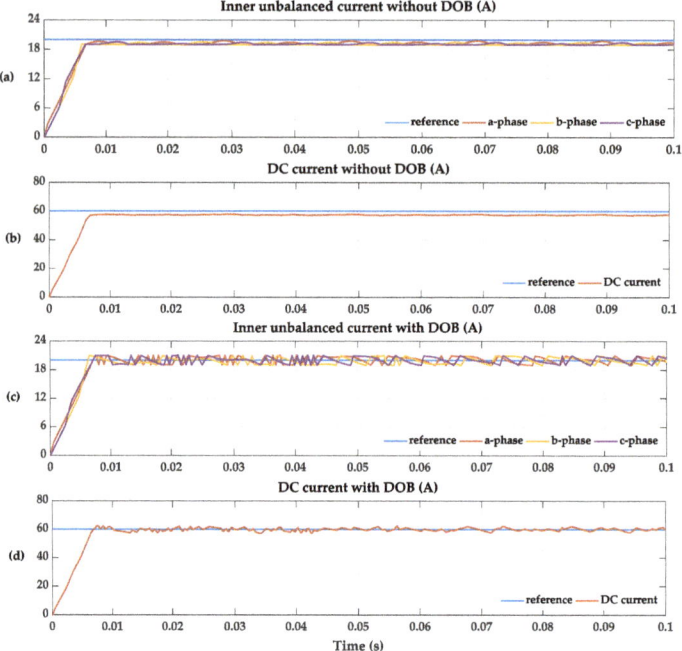

Figure 13. Simulations under actual inductance value increase condition: (**a**) Tracking of the inner unbalanced current to its reference without DOB; (**b**) Tracking of the DC current to its reference without DOB; (**c**) Tracking of the inner unbalanced current to its reference with DOB; (**d**) Tracking of the DC current to its reference with DOB.

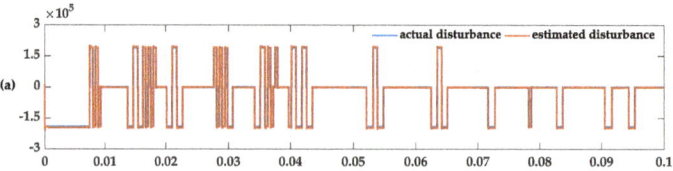

Figure 14. *Cont.*

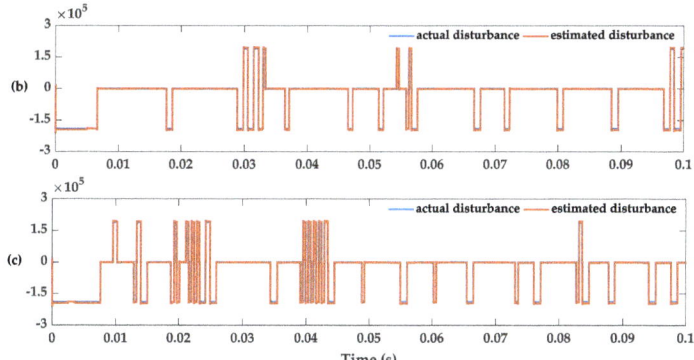

Figure 14. Actual disturbance value $d_{kn}^{(2)}$ and estimated disturbance value $\hat{d}_{kn}^{(2)}$ under actual inductance value increase condition: (**a**) A-phase; (**b**) B-phase; (**c**) C-phase.

4.5. Power Reversal

The battery energy storage system not only delivers energy to the grid through discharge, but also needs to get energy from the grid to recharge. Thus, the power reversal is simulated to verify the dynamic characteristics of the proposed control scheme. In this case, the AC reactive current reference is 0 A, the active current reference changes from 100 A to −100 A in the period of 0.05 s–0.1 s. The actual inductance values reduce by one tenth, the grid voltage contains 30% of 5th and 7th harmonics and the amplitude drops by 80%.

Figure 15a shows the DC current waveform. The batteries discharge before 0.075 s and the power reverses from 0.05 s to 0.1 s, then the batteries recharge until 0.15 s. Figure 15b represents the grid voltage with harmonics and sag. Aside from containing a lot of harmonics and ripples as seen in Figure 15c, the grid current is higher than its reference before 0.05 s and lower than its reference after 0.1 s. Figure 15d shows the improved current with DOB. The current keeps up with the reference stably. The tracing results of estimated disturbance value is presented in Figure 16. From these results, it is confirmed that the proposed scheme operates stably during dynamic periods.

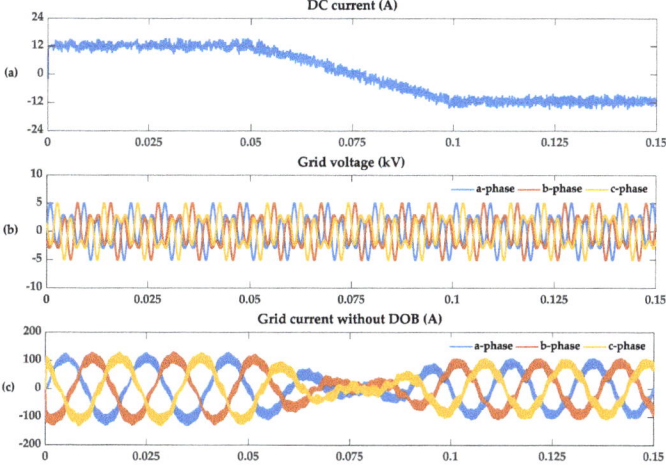

Figure 15. *Cont.*

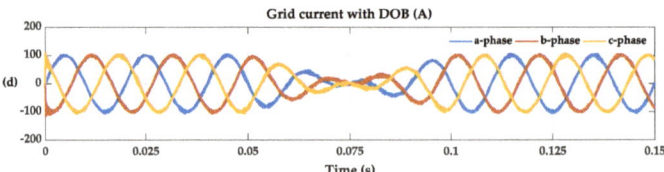

Figure 15. Simulations under power reversal condition: (**a**) DC current; (**b**) Grid voltage; (**c**) Grid current without DOB; (**d**) Grid current with DOB.

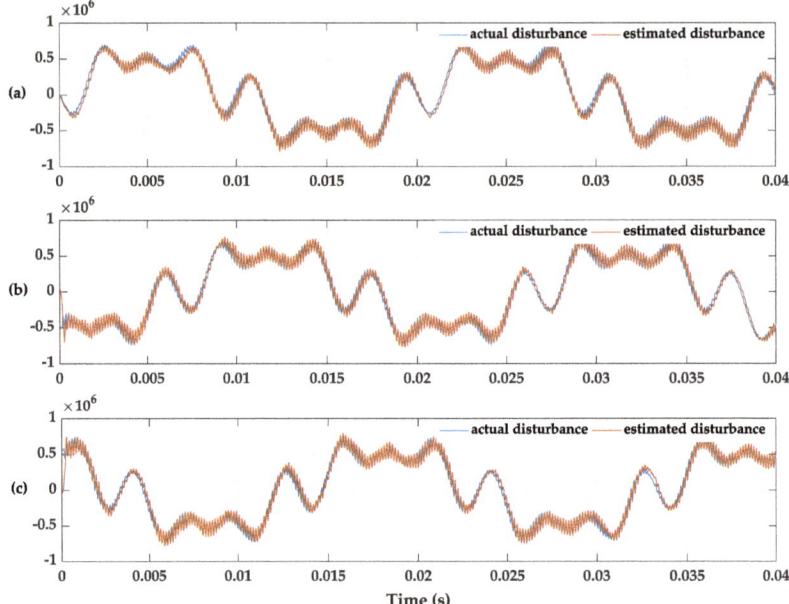

Figure 16. Actual disturbance value $d_{kn}^{(1)}$ and estimated disturbance value $\hat{d}_{kn}^{(1)}$ under power reversal condition: (**a**) A-phase; (**b**) B-phase; (**c**) C-phase.

5. Conclusions

A DOB-based MPC control scheme to restrain the disturbances caused by power quality problems and parameter mismatches is proposed. A MPC with less states and low calculation complexity is adopted and two linear discrete-time DOBs are designed. In addition, the method of parameter calculation is given. The proposed method redesigns the AC current control and circulating current control by combining with DOBs. The purpose is to enhance the anti-disturbance ability of MPC. The proposed DOB-based MPC control strategy has quite low cost of computation with the minimum order and optimal performance property.

The performances of the designed controller are simulated under five different system operations, including harmonics, 3-phase unbalance, voltage sag, parameter mismatches and power reversal. By comparing the control effect of AC current and circulating current under the original MPC controller and the MPC controller with DOB, respectively, the proposed method is validated as being effective. The AC current can maintain a low harmonic content and normal amplitude when it is disturbed by grid voltage, inductance and power. The inner unbalanced current and DC current can track the set-value stably under the inductance value increase condition. Besides, it is proved that the DOB is fast and steady by estimating the disturbances and uncertainties. However, a rigorous analysis of SM capacitor voltage balancing control is nontrivial, and further researches will be conducted to work on this interesting issue.

Author Contributions: All authors contributed collectively to the manuscript preparation and approved the final manuscript.

Funding: This research was funded by [State Grid Corporation of China science and technology project: Key technologies and equipment of DC step-up power collection and integration for large-scale PV Station].

Conflicts of Interest: The authors declare no conflict of interest.

References

1. Yang, T.; Mok, K.T.; Tan, S.C.; Lee, C.K.; Hui, S.Y.R. Electric springs with coordinated battery management for reducing voltage and frequency fluctuations in microgrids. *IEEE Trans. Smart Grid* **2018**, *9*, 1943–1952. [CrossRef]
2. Jiang, W.; Xue, S.; Zhang, L.; Xu, W.; Yu, K.; Chen, W.; Zhang, L. Flexible power distribution control in asymmetrical cascaded multilevel converter based hybrid energy storage system. *IEEE Trans. Ind. Electron.* **2018**, *65*, 6150–6159. [CrossRef]
3. Yang, P.; Xia, Y.; Yu, M.; Wei, W.; Peng, Y. A decentralized coordination control method for parallel bidirectional power converters in a hybrid AC/DC microgrid. *IEEE Trans. Ind. Electron.* **2018**, *65*, 6217–6228. [CrossRef]
4. Kwon, O.; Kim, J.S.; Kwon, J.M.; Kwon, B.H. Bidirectional grid-connected single-power-conversion converter with low input battery voltage. *IEEE Trans. Ind. Electron.* **2018**, *65*, 3136–3144. [CrossRef]
5. Mehrasa, M.; Pouresmaeil, E.; Zabihi, S.; Vechiu, I.; Catalão, J.P.S. A multi-loop control technique for the stable operation of modular multilevel converters in HVDC transmission systems. *Electr. Power Energy Syst.* **2018**, *96*, 194–207. [CrossRef]
6. Lee, K.J.; Park, B.G.; Kim, R.Y.; Hyun, D.S. Robust predictive current controller based on a disturbance estimator in a three-phase grid-connected inverter. *IEEE Trans. Power Electron.* **2012**, *27*, 276–283. [CrossRef]
7. Xia, C.; Wang, M.; Song, Z.; Liu, T. Robust model predictive current control of three-phase voltage source PWM rectifier with online disturbance observation. *IEEE Trans. Ind. Inform.* **2012**, *8*, 459–471. [CrossRef]
8. Mehrasa, M.; Pouresmaeil, E.; Akorede, M.F.; Zabihi, S.; Catalão, J.P.S. Function-based modulation control for modular multilevel converters under varying loading and parameters conditions. *IET Gener. Trans. Distrib.* **2017**, *11*, 3222–3230. [CrossRef]
9. Hagiwara, M.; Akagi, H. Experiment and simulation of a modular push–pull PWM converter for a battery energy storage system. *IEEE Trans. Ind. Appl.* **2014**, *50*, 1131–1140. [CrossRef]
10. Maharjan, L.; Yamagishi, T.; Akagi, H. Active-power control of individual converter cells for a battery energy storage system based on a multilevel cascade PWM converter. *IEEE Trans. Power Electron.* **2012**, *27*, 1099–1107. [CrossRef]
11. Song, J.; Zhang, W.; Liang, H.; Jiang, J.; Yu, W. Fault-tolerant control for a flexible group battery energy storage system based on cascaded multilevel converters. *Energies* **2018**, *11*, 171. [CrossRef]
12. Liang, H.; Guo, L.; Song, J.; Yang, Y.; Zhang, W.; Qi, H. State-of-charge balancing control of a modular multilevel converter with an integrated battery energy storage. *Energies* **2018**, *11*, 873. [CrossRef]
13. Ota, J.I.Y.; Sato, T.; Akagi, H. Enhancement of performance, availability, and flexibility of a battery energy storage system based on a modular multilevel cascaded converter (MMCC-SSBC). *IEEE Trans. Power Electron.* **2016**, *31*, 2791–2799. [CrossRef]
14. Zhang, M.; Huang, L.; Yao, W.; Lu, Z. Circulating harmonic current elimination of a CPS-PWM-based modular multilevel converter with a plug-in repetitive controller. *IEEE Trans. Power Electron.* **2014**, *29*, 2083–2097. [CrossRef]
15. Yang, L.; Li, Y.; Li, Z.; Wang, P.; Xu, S.; Gou, R. Loss optimization of MMC by second-order harmonic circulating current injection. *IEEE Trans. Power Electron.* **2018**, *33*, 5739–5753. [CrossRef]
16. Mehrasa, M.; Pouresmaeil, E.; Zabihi, S.; Catalao, J.P. Dynamic model, control and stability analysis of MMC in HVDC transmission systems. *IEEE Trans. Power Deliv.* **2017**, *32*, 1471–1482. [CrossRef]

17. Mehrasa, M.; Pouresmaeil, E.; Taheri, S.; Vechiu, I.; Catalão, J.P. Novel control strategy for modular multilevel converters based on differential flatness theory. *IEEE J. Emerg. Sel. Top. Power Electron.* **2017**, *6*, 888–897. [CrossRef]
18. Mehrasa, M.; Pouresmaeil, E.; Zabihi, S.; Caballero, J.T.; Catalão, J. A novel modulation function-based control of modular multilevel converters for high voltage direct current transmission systems. *Energies* **2016**, *9*, 867. [CrossRef]
19. Qin, J.; Saeedifard, M. Predictive control of a modular multilevel converter for a back-to-back HVDC system. *IEEE Trans. Power Deliv.* **2012**, *27*, 1538–1547.
20. Yang, J.; Li, S.; Chen, X.; Li, Q. Disturbance rejection of dead-time processes using disturbance observer and model predictive control. *Chem. Eng. Res. Des.* **2011**, *89*, 125–135. [CrossRef]
21. Qin, S.J.; Badgwell, T.A. A survey of industrial model predictive control technology. *Control Eng. Pract.* **2003**, *11*, 733–764. [CrossRef]
22. Liao, K.; Xu, Y. A robust load frequency control scheme for power systems based on second-order sliding mode and extended disturbance observer. *IEEE Trans. Ind. Inform.* **2018**, *14*, 3076–3086. [CrossRef]
23. Wang, C.; Mi, Y.; Fu, Y.; Wang, P. Frequency control of an isolated micro-grid using double sliding mode controllers and disturbance observer. *IEEE Trans. Smart Grid* **2018**, *9*, 923–930. [CrossRef]
24. Wang, B.; Dong, Z.; Yu, Y.; Wang, G.; Xu, D. Static-errorless deadbeat predictive current control using second-order sliding-mode disturbance observer for induction machine drives. *IEEE Trans. Power Electron.* **2018**, *33*, 2395–2403. [CrossRef]
25. Yang, J.; Cui, H.; Li, S.; Zolotas, A. Optimized active disturbance rejection control for DC-DC buck converters with uncertainties using a reduced-order GPI observer. *IEEE Trans. Circuits Syst.* **2018**, *65*, 832–841. [CrossRef]
26. Back, J.; Shim, H. Adding robustness to nominal output-feedback controllers for uncertain nonlinear systems: a nonlinear version of disturbance observer. *Automatica* **2008**, *44*, 2528–2537. [CrossRef]
27. Liu, H.; Li, S. Speed control for PMSM servo system using predictive functional control and extended state observer. *IEEE Trans. Ind. Electron.* **2011**, *59*, 1171–1183. [CrossRef]
28. Barros, J.D.; Silva, J.F.A.; Élvio, G.A.J. Fast-predictive optimal control of NPC multilevel converters. *IEEE Trans. Ind. Electron.* **2012**, *60*, 619–627. [CrossRef]
29. Nguyen, H.T.; Jung, J.W. Disturbance-rejection-based model predictive control: flexible-mode design with a modulator for three-phase inverters. *IEEE Trans. Ind. Electron.* **2018**, *65*, 2893–2903. [CrossRef]
30. Moon, J.W.; Gwon, J.S.; Park, J.W.; Kang, D.W.; Kim, J.M. Model predictive control with a reduced number of considered states in a modular multilevel converter for HVDC system. *IEEE Trans. Power Deliv.* **2015**, *30*, 608–617. [CrossRef]
31. Tu, Q.; Xu, Z.; Xu, L. Reduced switching-frequency modulation and circulating current suppression for modular multilevel converters. *IEEE Trans. Power Deliv.* **2011**, *26*, 2009–2017.
32. Moon, J.W.; Kim, C.S.; Park, J.W.; Kang, D.W.; Kim, J.M. Circulating current control in MMC under the unbalanced voltage. *IEEE Trans. Power Deliv.* **2013**, *28*, 1952–1959. [CrossRef]
33. Kim, K.S.; Rew, K.H. Reduced order disturbance observer for discrete-time linear systems. *Automatica* **2013**, *49*, 968–975. [CrossRef]

© 2018 by the authors. Licensee MDPI, Basel, Switzerland. This article is an open access article distributed under the terms and conditions of the Creative Commons Attribution (CC BY) license (http://creativecommons.org/licenses/by/4.0/).

Article

A Centralized Smart Decision-Making Hierarchical Interactive Architecture for Multiple Home Microgrids in Retail Electricity Market

Masoumeh Javadi [1,2], Mousa Marzband [3,4], Mudathir Funsho Akorede [5], Radu Godina [6,*], Ameena Saad Al-Sumaiti [7] and Edris Pouresmaeil [8]

1. Department of Electrical Power Engineering, Guilan Science and Research Branch, Islamic Azad University, Rasht 4147654919, Iran; javadi.masoomeh@gmail.com
2. Department of Electrical Power Engineering, Rasht Branch, Islamic Azad University, Rasht 4147654919, Iran
3. Faculty of Engineering and Environment, Department of Maths, Physics and Electrical Engineering, Northumbria University Newcastle, Newcastle upon Tyne NE1 8ST, UK; mousa.marzband@northumbria.ac.uk
4. Department of Electrical Engineering, Lahijan Branch, Islamic Azad University, Lahijan 4416939515, Iran
5. Department of Electrical & Electronics Engineering, Faculty of Engineering and Technology, University of Ilorin, P.M.B. 1515 Ilorin, Nigeria; mudathir.akorede@yahoo.com
6. Centre for Aerospace Science and Technologies—Department of Electromechanical Engineering, University of Beira Interior, 6201-001 Covilhã, Portugal
7. Electrical and Computer Engineering, Khalifa University, Abu Dhabi 127788, UAE; ameena.alsumaiti@ku.ac.ae
8. Department of Electrical Engineering and Automation, Aalto University, 02150 Espoo, Finland; edris.pouresmaeil@gmail.com
* Correspondence: rd@ubi.pt; Tel.: +351-96-440-2819

Received: 3 October 2018; Accepted: 10 November 2018; Published: 14 November 2018

Abstract: The principal aim of this study is to devise a combined market operator and a distribution network operator structure for multiple home-microgrids (MH-MGs) connected to an upstream grid. Here, there are three distinct types of players with opposite intentions that can participate as a consumer and/or prosumer (as a buyer or seller) in the market. All players that are price makers can compete with each other to obtain much more possible profitability while consumers aim to minimize the market-clearing price. For modeling the interactions among partakers and implementing this comprehensive structure, a multi-objective function problem is solved by using a static, non-cooperative game theory. The propounded structure is a hierarchical bi-level controller, and its accomplishment in the optimal control of MH-MGs with distributed energy resources has been evaluated. The outcome of this algorithm provides the best and most suitable power allocation among different players in the market while satisfying each player's goals. Furthermore, the amount of profit gained by each player is ascertained. Simulation results demonstrate 169% increase in the total payoff compared to the imperialist competition algorithm. This percentage proves the effectiveness, extensibility and flexibility of the presented approach in encouraging participants to join the market and boost their profits.

Keywords: demand side management; electricity market; game theory; home energy management system; home microgrid; Nikaido-Isoda function

1. Introduction

A home microgrid (H-MG) consists of locally distributed energy resources (DERs) which comprise non-dispatchable renewable energy resources, dispatchable resources, energy storage (ES) and responsive load demand (RLD). It can either supply its local loads independently or connected to the upstream grid. For an optimal use of the DERs present in a H-MG, the mismatch of power between the energy production and consumption must be reduced to the barest minimum [1]. Operating a H-MG optimally would not only contribute to the electric utility profit [2], but also would improve the reliability of the system besides the proper load distribution management [3]. Indeed, the optimum exploitation of H-MGs has become an important topic necessitating further research to achieve better operation of their hybrid energy resources and also their demand management. Hence, hierarchical domination structures have been executed to guarantee a dependable performance of the power flow among DC H-MG groups in a neighborhood system [4] and also connect to an AC bus to adjust the system stability [5]. However, during implementing the economic dispatch in H-MGs, decisions of an energy management system (EMS) could be affected by DER, ES and RLD bids [6]. In addition, achieving a dynamic exploitation and control plans for a hybrid H-MG can assist in providing reactive power and set the voltage [7,8] in order to solve problems of power stability like oscillations in a hybrid multi-system [9], asymmetrical faults [10] and ground fault [11]. Designing an efficient EMS at the residential level depends heavily on the electricity price and necessitates the consideration of households patterns [12]. Furthermore, the remarkable participation of householders whose houses are equipped with renewable energy resources and ES in demand response programs, while reducing carbon emissions would make an impact on the market as they are to reduce their cost [13]. So, for expanding these participators in demand-side management programs, the EMS needs an interactive and user-friendly interface with secure communication [14]. Moreover, H-MGs should be armed with a decision support tool for adopting their initial strategies [15] based on local optimization of DER operation and energy usage by a domestic energy management controller [16] that enable them to engage in the market eagerly.

Consequently, one of the benefits of operating multiple home-microgrid (MH-MG) systems is the concurrent operation and the optimum use of DERs existing in each H-MG. This implies strategies for storing energy in a H-MG during excess generation in other H-MGs and/or supplying the required demand of the H-MG that cannot meet its power demand. In other words, H-MGs can play the role of both a generating player and a consuming player during the time period [17]. Hence, a H-MG could either meet its demand from the energy produced by itself or seek aid from other H-MGs [18]. The H-MG with excess generation (generating player) must supply its power to H-MGs having a power shortage (consuming player) and/or to the upstream grid.

To attain this goal, tools such as the active response of consumers to the demand [19], the implementation of a powerful EMS [20], and the adequate power dispatch in smart grids are required. One of the challenges in this regard is the coordination between energy management functions, having concentrated control or hierarchical systems inside H-MGs [21]. Another challenge is the selection of an adequate formula for the optimization problem considering the keen competition between partakers with contradictory intentions. Although an economic dispatch of DERs in a MH-MG system through applying the Carnot model has been presented in [18], a multi-objective function for reaching the collective payoff within competition between diverse players under game theory procedure has not been studied. Furthermore, reference [22] presented a statical optimization formula but the distributed storage system was not considered in that study. However, dynamic energy storage systems models for the overall energy management of H-MGs were proposed in [23].

In the same vein, a parallel exploitation of H-MGs was investigated in [24], just as the technical frameworks and the economic aspects of this structure was presented in [25]. Results of the investigations showed that in addition to creating a competitive market, the back-to-back connections among H-MGs could provide much better separation and load distribution control relative to the single H-MG structure.

H-MGs receiving power supply and other H-MGs must have the possibility of using an upstream grid, the non-dispatchable units (NDUs) and their local energy storage systems. In addition, H-MGs ought to have access to other neighboring H-MGs for swapping excess electricity while all types of players are satisfied with the best possible way they achieve their defined objective functions. This is the duty of the EMS as it satisfies technical and economic constraints related to each generation and consumption to establish the best choice for power equilibrium in the network [26]. In a system of MH-MGs, the distribution networks (market operator (MO) and a distribution network operator (DNO)), and H-MGs desire the optimum utilization of power generation and consumption resources. To attain higher reliability, the EMSs must have this capability to store the maximum possible energy in ESs of each H-MG. From this viewpoint, the optimum design of a system with MH-MG leads to the coincident optimization of H-MGs and distribution network pay-offs. Such a design has a dynamic programming nature.

In this paper, a retail market optimization structure for multi-ownership systems with MH-MG including players with opposite goals is recommended. Using a non-cooperative game theory approach based on the supply function model assists in analyzing the electricity buyers and sellers' individual behaviors in the market by enabling a competition in energy trading between H-MGs. Therefore, an active distributed system through the presented method will be provided. The proposed structure can handle the interconnection of MH-MGs with various DER resources capacities and the independent and communal performance of each H-MG. Indeed, this structure is comprehensive in its capability of accepting any DER technology and the participation of distribution companies (retailers) in the market structure. The proposed structure is advantageous as it can improve the economic productivity of the participating H-MGs.

The significant contributions of this study can be described as follows:

- Persuading further residential users to be equiped with DERs and ES in order to be involved in energy trading and RLD management program;
- Proposing a retail competition market model to trade distributed energy by ensuring fairness among non-cooperative players through a stochastic, and autonomous decision-making structure;
- Enhancing the economic operation and profitably of all members (about 169% boost in the collective payoff compared with the imperialist competition algorithm (ICA) results [18]).

The remainder of the paper is organized as follows: The overview of the MH-MG concept is provided in Section 2. The general outline of the network under study is presented in Section 3. The proposed market structure is described in Section 4. The market optimization problem formulation is explained in Section 5. The procedure of implementing the Nikaido-Isoda/relaxation algorithm (NIRA) is stated in Section 6. Simulation results and discussion of the proposed case study is validated in Section 7. The conclusion is given in Section 8.

2. MH-MG Concept

The MH-MG system in this paper, shown in Figure 1, refers to a network of H-MGs that swap electricity with each other to supply their neighbors' shortage whereas trying to maximize their own payoffs. Indeed, each individual H-MG is like a green building that consists of local generation resources, ES devices and loads. Similar to conventional MGs, green buildings are able to autonomously support their demand to some extent [27]. These kinds of buildings possess the ability to act as a generation, storage, and demand response unit, in a similar manner to a MG. Also, green buildings can take part in a retail market to trade energy with other green buildings [18]. Since the main focus of this study is on managing local energy networks of a residential district, the concept of MH-MG has been used in this paper as in other research at the residential level [3,28–30].

In fact, similar to interoperability of multiple MGs in an integrated system, H-MGs with excess generation are able to supply other H-MGs' needs that a face power shortage. Thus, some H-MGs act as generators for maximizing their profits resulting from selling energy to the market and others act as consumers for reducing electricity price through demand-side management. On the other hand, like multiple MG network, an EMS for monitoring players' strategies and therefore adopting fair decisions in energy trading is necessary. Therefore, a central energy management system (CEMS) is an essential element in the MH-MG network.

For participating in electricity deals in the market, each H-MG in this paper is formed of two players including a generator and a consumer. This characteristic is considered here in order to contribute to implementing the market structure related to every ownership condition. For instance, at a time when a H-MG's tenants are not the possessor of the building, DERs or ES devices, the owner of building is the generating player and the tenant is the consuming player. In this case, the formulation of players' tactics in a competitive situation is conveniently possible.

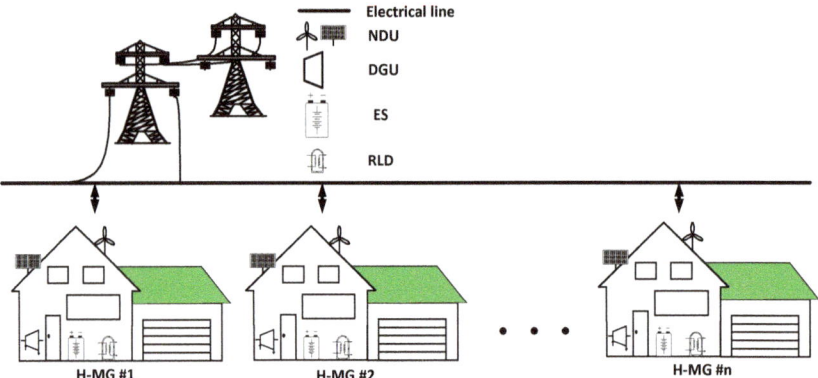

Figure 1. A multiple home-microgrid (MH-MG) system. Non-dispatchable unit (NDU), dispatchable generation unit (DGU), energy storage (ES), responsive load demand (RLD).

3. General Outline of the Network under Study

A network structure with MH-MG, retailers, a MO and a DNO is proposed as shown in Figure 2. The MH-MGs interact with each other and with retailers for the exchange of power and the optimal utilization of power generation resources. The MO proposes the optimum price upon receiving price suggestions from buyers and sellers and the execution of power dispatch by the DNO. Although the DNO is the owner and exploiter of the equipment and distribution network cables, it is not involved in the act of selling of electricity.

Each H-MG includes non-responsive loads (NRL) and DERs that comprise RLD, ES resources, controllable generation resources and non-controllable generation resources. DERs are grouped into generating players while the consumed resources (i.e., RLD) in each H-MG are grouped as consuming players. Each group is to target an objective function. The power producing (generating) players are to maximize their profit. In comparison, the consuming players are to minimize their cost.

According to the priority included based on the price suggestions of H-MGs, each MH-MG has the duty at the beginning to supply local loads through generation resources. During each time interval, H-MGs may encounter a power generation shortage and/or an excess power generation depending on the amount of power produced by each MH-MG and/or the amount of their local load demand. On the other hand, when each H-MG encounters an excess generation, it tends to sell its power at a higher price to distribution companies or other H-MGs. In other words, if a H-MG encounters a power shortage, it compensates for that by setting a price lower than other alternatives. Therefore, each player must perform a comparison analysis between the proposed prices by other H-MGs and distribution companies for the selection of the optimal price.

Each MH-MG participates with its suggested price in this proposed market which may fail in its excess power transactions due to their higher bids. To encourage further participation of H-MGs in this process, the distribution companies buy the amount of excess generation of each H-MG that has not succeeded in selling to other H-MGs. In addition, power equilibrium is also established in each H-MG and the power network.

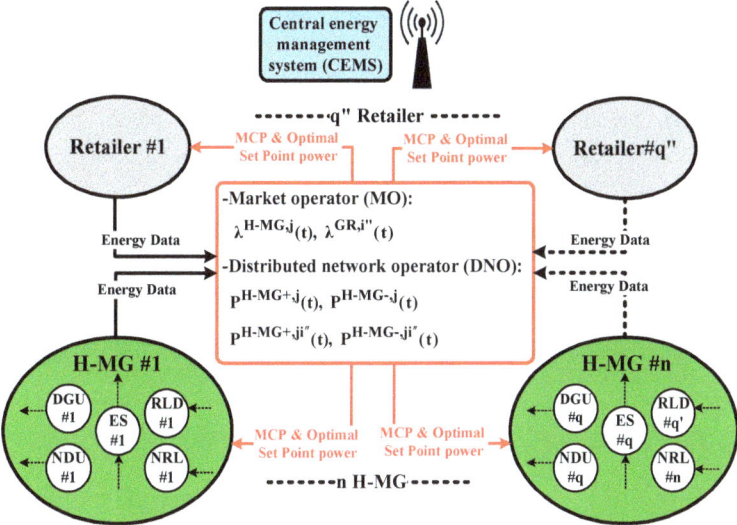

Figure 2. Interaction of distributed network operator (DNO), market operator (MO) and MH-MGs. Central energy management system (CEMS), non-responsive load (NRL), market clearing price (MCP).

4. The Proposed Market Structure

The proposed retail electricity market structure presents a solution for providing distribution generators with large portions of their capacities to participate in the market. It reduces the electricity price thereby increasing profit alongside their effective and efficient interaction with consumers.

The framework considered in this work provides the exploiters of distribution system and domestic customers with this possibility of properly selecting their energy supply source considering various options such as choosing a comprehensive range of renewable energy resources based on the market clearing price. The recommended market structure is presented in Figure 3. The following stages describe the market operation.

Stage 1   In the first stage, the prediction data of NDU and the consumed load of MH-MG are entered into the scenario generation phase.

Stage 2   Next, stage 2 is focused on generating uncertainty scenarios considering the prediction data of stage 1 with the corresponding occurrence probability. Also in this stage, the participation of generating units and consumers is planned proportionally to the generated scenarios in each MH-MG. Moreover, the optimum programming is handled in this stage based on the units' participation price (price-based unit commitment) in order to determine the maximum available capacities of players for engaging in the market.

Stage 3   The third stage is to calculate the expected value (EV) of random quantities related to uncertainty scenarios of players for participating in game theory and determining the Nash equilibrium (participation optimum capacity) in market clearing price with random optimization approach based on calculating the value of Nikaido-Isoda function and relaxation algorithm.

Stage 4   The final stage is for determining the optimum capacity of the players for participating in the market and calculating the payoff function of each one of them.

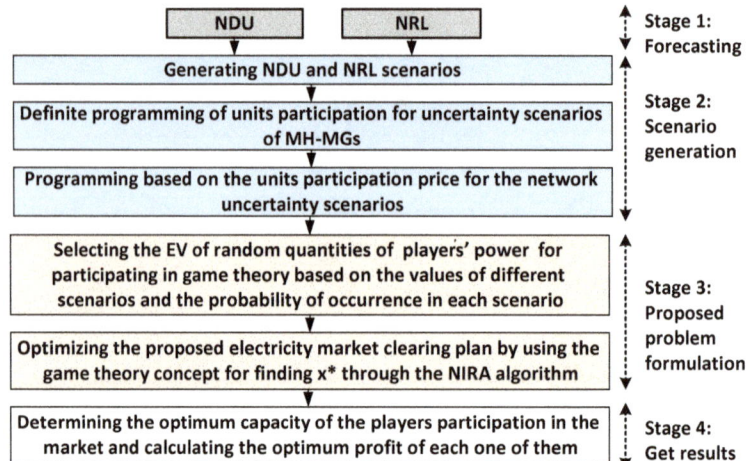

Figure 3. The process of implementing the proposed market structure. Expected value (EV), Nikaido-Isoda/relaxation algorithm (NIRA).

5. The Market Optimization Problem Formulation

The major elements of the proposed market structure include distribution companies and H-MGs of two players (prosumers and consumers). The mathematical model including the objective function and constraints for each category will be explored in this section.

Objective Functions and Problem Constraints

The main elements of the proposed market structure include distribution companies and H-MGs consisting of two players which includes generation and consumption. The objective functions for each one of them can be defined as follows:

- **Power Generation Unit**

The power generation resources in the studied MH-MGs are dispatchable generation units (DGUs), NDU and ES. The objective function is to maximize the profit obtained from a generator #i at time t as defined by ($\mathbb{J}^i(t)$) in Equation (1)

$$\max \ \mathbb{J}^i(t) = \mathbb{R}^i(t) - \mathbb{C}^i(t), \ t \in \{1, 2, \cdots, 24\}, \ i \in \{1, 2, \cdots, q\} \tag{1}$$

$$\mathbb{R}^i(t) = \lambda^{\text{H-MG},j}(t) \times [P^{\text{DGU},j}(t) + P^{\text{NDU},j}(t) + P^{\text{ES}-,j}(t) - P^{\text{NRL},j}(t)], \ j \in \{1, 2, \cdots, n\} \tag{2}$$

For comprehensibility, the retail electricity price for all players in an H-MG is presumed the same. Therefore, following relations apply.

$$\lambda^{\text{H-MG},j}(t) = (-\theta \times P^{\text{NRL},j}(t)) + \beta, \ \theta > 0 \tag{3}$$

$$\mathbb{C}^i(t) = \mathbb{C}^{\text{DGU},j}(t) + \mathbb{C}^{\text{NDU},j}(t) + \mathbb{C}^{\text{ES}-,j}(t) + \mathbb{C}^{\text{ES}+,j}(t) + \mathbb{C}^{\text{H-MG}+,j}(t) \tag{4}$$

$$\mathbb{C}^{\text{DGU},j}(t) = a^j \cdot (P^{\text{DGU},j}(t))^2 + b^j \cdot P^{\text{DGU},j}(t) + c^j, \ a^j > 0 \tag{5}$$

$$\mathbb{C}^{\text{ES}-,j}(t) = \pi^{\text{ES}-} \times P^{\text{ES}-,j}(t), \ \mathbb{C}^{\text{ES}+,j}(t) = \pi^{\text{ES}+} \times P^{\text{ES}+,j}(t) \tag{6}$$

Should any H-MG face a shortage in satisfying the needs of RLD and NRL loads of its MH-MG, it must compensate the power shortage by buying power from other H-MGs and/or the network by selecting the least cost offer. Thus, $\mathbb{C}^{\text{H-MG}+,j}(t)$ can be computed as follows:

$$\mathbb{C}^{\text{H-MG+},j}(t) = \mathbb{C}^{\text{H-MG+},jm}(t)|_{m \neq j} + \mathbb{C}^{\text{H-MG+},ji''}(t) \tag{7}$$

$$\mathbb{C}^{\text{H-MG+},ji''}(t) = (1 - X^{\text{H-MG+},j}(t)) \times ((P^{\text{H-MG+},j}(t) - \sum_{m=1}^{n} (1 - X^{\text{H-MG},m}(t)) \times P^{\text{H-MG-},m}(t)) \times \lambda^{\text{GR},i''}(t)) \tag{8}$$

The above expressions are such that its shortage is compensated by comparing the prices and power exchange capacity of other H-MGs. In case the power required by the H-MG #j is not satisfied through the power exchange with other H-MGs (see (9)), the H-MG will compensate the power deficit by buying power from distribution networks. The intention of H-MGs is to minimize the buying cost while satisfying their load demand. Such a goal is made possible by comparing the offer of other H-MGs to that of the distribution grid (i.e., $\lambda^{\text{GR},i''}(t)$)

$$X^{H-MG,j}(t) = [X^{\text{H-MG},1}(t), X^{\text{H-MG},2}(t), \cdots, X^{\text{H-MG},n}(t)] \tag{9}$$

The surplus and scarcity of power related to each H-MG is stored in a variable as follows:

$$P^{\text{H-MG},j}(t) = [P^{\text{H-MG},1}(t), P^{\text{H-MG},2}(t), \cdots, P^{\text{H-MG},n}(t)] \tag{10}$$

The offer by each H-MG can also be stored in the following variable:

$$\lambda^{\text{H-MG},j}(t) = [\lambda^{\text{H-MG},1}(t), \lambda^{\text{H-MG},2}(t), \cdots, \lambda^{\text{H-MG},n}(t)] \tag{11}$$

The information related to a tertiary block during each time interval in a matrix is stored as follows:

$$\Omega^{\text{H-MG},j}(t) = \begin{bmatrix} \lambda^{\text{H-MG},1}(t) & \lambda^{\text{H-MG},2}(t) & \cdots & \lambda^{\text{H-MG},n}(t) \\ P^{\text{H-MG},1}(t) & P^{\text{H-MG},2}(t) & \cdots & P^{\text{H-MG},n}(t) \\ X^{\text{H-MG},1}(t) & X^{\text{H-MG},2}(t) & \cdots & X^{\text{H-MG},n}(t) \end{bmatrix} \tag{12}$$

The $\Omega^{\text{H-MG},j}(t)$ variable proportional to the offer of each H-MG arranged in ascending order, is defined as follows:

$$\Omega'^{\text{H-MG},j}(t) = \begin{bmatrix} \lambda'^{\text{H-MG},1}(t) & \lambda'^{\text{H-MG},2}(t) & \cdots & \lambda'^{\text{H-MG},n}(t) \\ P'^{\text{H-MG},1}(t) & P'^{\text{H-MG},2}(t) & \cdots & P'^{\text{H-MG},n}(t) \\ X'^{\text{H-MG},1}(t) & X'^{\text{H-MG},2}(t) & \cdots & X'^{\text{H-MG},n}(t) \end{bmatrix} \tag{13}$$

where $\lambda'^{\text{H-MG},1}(t) < \lambda'^{\text{H-MG},2}(t) < \cdots < \lambda'^{\text{H-MG},n}(t)$. The amount of power shortage of H-MG #j can be compensated by other H-MGs proportional to the order of their offer. So, this power shortage must be compared with the excess power generated by other resources and compensated accordingly. The possibility of supplying H-MG #j power shortage through the excess power generated by other H-MGs causes the binary variable matrix condition change. This is indicated by $X''^{\text{H-MG}}(t)$ in (14). The component proportional to this matrix becomes one for a total or a partial supply.

$$X''^{\text{H-MG},j}(t) = [X''^{\text{H-MG},1}(t), X''^{\text{H-MG},2}(t), \cdots, X''^{\text{H-MG},n}(t)]_{n \neq j} \tag{14}$$

The least buying cost that H-MG #j bears if encountering a power shortage is computed by (15, 16).

$$\mathbb{C}^{\text{H-MG+},jm}(t) = X''^{\text{H-MG},j}(t) \times \lambda^{\text{H-MG},j}(t) \times \Delta P \tag{15}$$

$$\Delta P = (P^{\text{H-MG},j}(t) - P^{\text{H-MG},m}(t))_{j \neq m} \tag{16}$$

- **Consumers**

Consumers are a sort of players with RLD loads in each MH-MG. The aim of this group is to minimize the exploitation cost by managing their distributable loads as represented by the objective function in (17).

$$\min \; \mathbb{J}^{\prime i^\prime}(t) = \lambda^{\text{H-MG},j}(t) \times P^{\text{RLD},j}(t), \; i^\prime \in \{1,2,\cdots,q^\prime\} \tag{17}$$

- **Upstream Grid**

This collection includes the amount of participation of distribution networks in buying the surplus power from H-MGs and also vending power to H-MGs in the case of a lack of power. $\mathbb{J}^{\prime\prime\text{GR},i^{\prime\prime}}(t)$ is defined as the earnings obtained from swapping the distribution network power at time t. The objective is to maximize it as shown below:

$$\max \; \mathbb{J}^{\prime\prime\text{GR},i^{\prime\prime}}(t) = \mathbb{R}^{\text{GR},i^{\prime\prime}}(t) - \mathbb{C}^{\text{GR},i^{\prime\prime}}(t), \; i^{\prime\prime} \in \{1,2,\cdots,q^{\prime\prime}\} \tag{18}$$

$$\mathbb{R}^{\text{GR},i^{\prime\prime}}(t) = \lambda^{\text{GR},i^{\prime\prime}}(t) \times \sum_{j=1}^{n} P^{\text{H-MG+},ji^{\prime\prime}}(t), \; \mathbb{C}^{\text{GR},i^{\prime\prime}}(t) = \sum_{j=1}^{n} \lambda^{\text{H-MG},j}(t) \times P^{\text{H-MG-},ji^{\prime\prime}}(t) \tag{19}$$

- **Operational Constraints**

The operation of players and the system is subject to a variety of constraints. These constraints include power balance constraint (20), the power generation limits on the DGU (Equation (21)) and NDU (Equations (22) and (23)), the ES charging/discharging constraints (Equations (24)–(26)) [3,31], RLD limits (Equation (27)) [3], and the power exchange between H-MGs constraint (Equations (28)–(30)). It is important to emphasize that ξ in (Equation (27)) shows that the value of RLD is considered as a part of NRL.

$$\sum_{j=1}^{n} P^{\text{DGU},j}(t) + P^{\text{NDU},j}(t) + P^{\text{ES-},j}(t) + P^{\text{H-MG+},ji^{\prime\prime}}(t)$$
$$= \sum_{j=1}^{n} P^{\text{NRL},j}(t) + P^{\text{ES+},j}(t) + P^{\text{RLD},j}(t) + P^{\text{H-MG-},ji^{\prime\prime}}(t) \tag{20}$$

$$\underline{P}^{\text{DGU},j} \leq P^{\text{DGU},j}(t) \leq \overline{P}^{\text{DGU},j}, \; \forall t \tag{21}$$

$$0 \leq P^{\text{NDU},j}(t) \leq EV^{\text{NDU},j}(t), \; \forall t \tag{22}$$

$$EV^{\text{NDU},j}(t) = \sum_{s=1}^{N_s} \rho_s^{\text{NDU},j}(t) \times P_s^{\text{NDU},j}(t) \tag{23}$$

$$0 \leq P^{\text{ES-},j}(t)(P^{\text{ES+},j}(t)) \leq \overline{P}^{\text{ES-},j}(\overline{P}^{\text{ES+},j}), \; \forall t \tag{24}$$

$$\underline{SOC}^{\text{ES},j} \leq SOC^{\text{ES},j}(t) \leq \overline{SOC}^{\text{ES},j} \tag{25}$$

$$SOC^{\text{ES},j}(t+1) - SOC^{\text{ES},j}(t) = \frac{(P^{\text{ES+},j}(t) - P^{\text{ES-},j}(t)) \times \Delta t}{ES_{\text{Tot}}^{\text{ES},j}} \tag{26}$$

$$0 \leq P^{\text{RLD},j}(t) \leq \xi \times P^{\text{NRL},j}(t) \tag{27}$$

$$0 \leq \sum_{j=1}^{n} P^{\text{H-MG+},ji^{\prime\prime}}(t)(\sum_{j=1}^{n} P^{\text{H-MG-},ji^{\prime\prime}}(t)) \leq EV^{\text{H-MG+},ji^{\prime\prime}}(t)(EV^{\text{H-MG-},ji^{\prime\prime}}(t)) \tag{28}$$

$$EV^{\text{H-MG+},ji^{\prime\prime}}(t) = \sum_{s=1}^{N_s} \rho_s^{\text{H-MG+},ji^{\prime\prime}}(t) \times P_s^{\text{H-MG+},ji^{\prime\prime}}(t) \tag{29}$$

$$EV^{\text{H-MG-},ji^{\prime\prime}}(t) = \sum_{s=1}^{N_s} \rho_s^{\text{H-MG-},ji^{\prime\prime}}(t) \times P_s^{\text{H-MG-},ji^{\prime\prime}}(t) \tag{30}$$

6. Implementing the NIRA Algorithm

A random early retail energy market-based on the Nikaido-Isoda/relaxation (REM-NIRA) algorithm is presented to provide a comprehensive and scalable solution where any number of players can take part in trading energy [32]. The Algorithm will be applied to find an electricity market equilibrium in order to clear the retail electricity market price through analyzing the players' behavior by using the concept of Nash equilibrium as a solution in the multi-agent interaction problems. A flowchart explaining the algorithm is presented in Figure 4. The flowchart consists of primary and secondary levels. A description of each level is provided in this section.

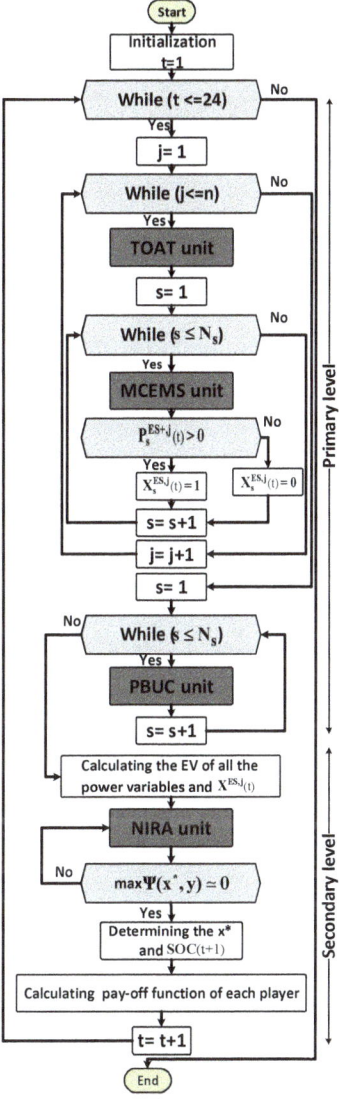

Figure 4. Flowchart of the proposed algorithm for implementing the retail energy market based on Nikaido-Isoda/relaxation algorithm (REM-NIRA). Taguchi's orthogonal array testing (TOAT) unit, modified conventional energy management system (MCEMS) unit, and price-based unit commitment (PBUC).

6.1. Primary Level of REM-NIRA Algorithm

The primary level of REM-NIRA algorithm is composed of three main units: the Taguchi's orthogonal array testing (TOAT) unit, the modified conventional energy management system (MCEMS) unit, and the price-based unit commitment (PBUC) unit. The primary level is to achieve the following tasks:

1. Determining the amount of power generated by all generation sources along with the corresponding probabilities of each power generation scenario;
2. Determining the power consumed by all RLD and NRL along with their corresponding probabilities of each demand scenario;
3. Estimating the amount of the deficiency and surplus of power related to each H-MG;
4. Defining the grid capacity in terms of power purchase and power sale.

TOAT is an approach which has been applied to choose minimum optimal representative scenarios. Moreover, for local scheduling of initial powers of H-MGs in the proposed structure, the MCEMS algorithm has been used. Since the operation of the TOAT unit and the MCEMS unit is explained in detail in [3,32], only a description of the PBUC unit will be discussed here.

The purpose of the PBUC unit is to establish the grid power set-point with generation resources and consumption of H-MGs. This unit encourages H-MGs to participate in a retail market while satisfying their needs. Taking into consideration the offer price of each H-MG and the grid, the capacity of the distribution network in terms of power purchase and sale uncertainty scenarios are to be determined. The structure of this unit is implemented according to Figure 5. The initial values of participation of grid variables for selling to and buying from H-MGs are determined based on players' accessible capacities and their bids for the NIRA unit.

6.2. Secondary Level of REM-NIRA Algorithm

The second level of the REM-NIRA algorithm structure consists of a main unit called the NIRA unit (the NIRA algorithm is explained in detail in [32]). The initial guess for the unit is chosen based on the data acquired from the primary level scenarios. In this regard, it is assumed that the nature of the discussed electricity market is proportionate to the game theory with n entrants in a non-cooperative game. In the unit, each player maximizes their benefit through a centralized decision making procedure. The objective of this level is to determine players' Nash equilibrium by utilizing the game theory specially designed means (NIRA algorithm). Having known the balanced response through continuous iterative loops, the electricity market price can be cleared for a MH-MG having several customers.

Through the NIRA unit, two coupled sub-problems are solved including: (1) Maximizing the Nikaido-Isoda function and (2) employing the relaxation algorithm and improving the optimal response function [32]. Both objectives are followed interactively by the NIRA unit until the contrast in the optimal response function between the two consecutive iterations becomes smaller than a predefined threshold. After the initial value definition and forming a pay-off function for each player based on such values, as well as forming a Nikaido-Isoda function at this level, the Nikaido-Isoda function must be maximized first. Then, gradually, the obtained solution from this function in the first sub-problem meets a new stable state showing the proper results.

After obtaining the intermediate solution in the first sub-problem, It is the second sub-problem's turn to run. In the second sub-problem, the relaxation algorithm is applied to improve the solution space and update it. If values of the Nikaido-Isoda function reach zero, no players can unilaterally improve their payoff function. Therefore, a balanced (approximate) response is found for the electricity market clearing by following the general and local constraints (Equations (20)–(30)). With the repeated improvement of the optimal response function, the values of the payoff function of all the players gradually converge to an equilibrium (approximate) point. The aim of implementing the secondary level is to attain the following:

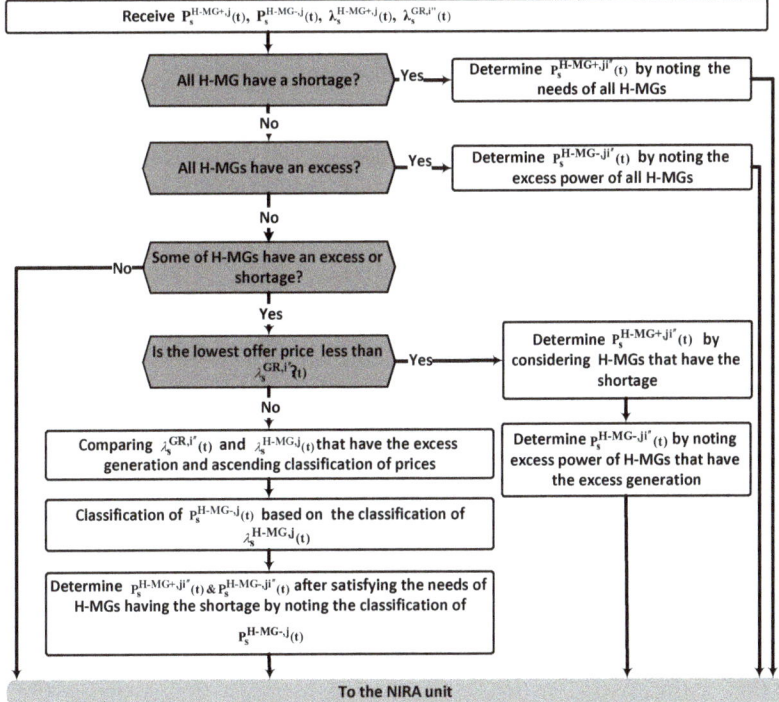

Figure 5. PBUC unit.

1. Initial guess based on players' EVs;
2. x^* vector (the optimum capacity of players' participation in the network) based on the Nash equilibrium of players;
3. The optimum amount of profit for players.

7. Simulation Results and Discussion

In order to test the capability of the proposed method for running the market, a case study has been developed in a MATLAB software simulation environment. The details of the entire system and the principles of the control plan for each of the DERs are presented in Appendix A. The predicted data of NRL, NDU (here, wind turbine and photo-voltaic panel) are taken from [18]. Figure 6 shows the configuration of the system under study which consists of MH-MGs and the network.

Each H-MG is an energy district consisting of a set of generation resources, which include NDU, DGU, ES, NRL, and RLD. The number of MH-MGs and the connected distribution networks are expanded to n and q'' values. For the system under study, three H-MGs and a distribution network are considered. To investigate the performance of the proposed REM-NIRA algorithm, the following scenarios are considered on the network case study:

Scenario #1: Normal operation.
Scenario #2: Sudden NDU generation increase (by 10%).
Scenario #3: Sudden NDU generation decrease (by 10%).

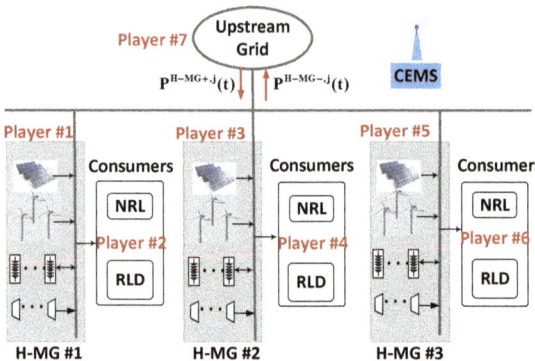

Figure 6. The network under study.

For all scenarios, the amount of produced power by NDUs of each H-MG, and also the amount of consumed NRL (after applying the uncertainty) during a day is shown in Figure 7. The peak power consumption of H-MGs is mainly in the early hours of the morning and night as seen in Figure 7. Although, during these hours, the load demand in all H-MGs is far greater than the amount of power that is generated by the NDUs, remaining demand can be met by other options such as the generated power by DGUs, controlling demand by the RLD program, or purchasing power from the upstream grid.

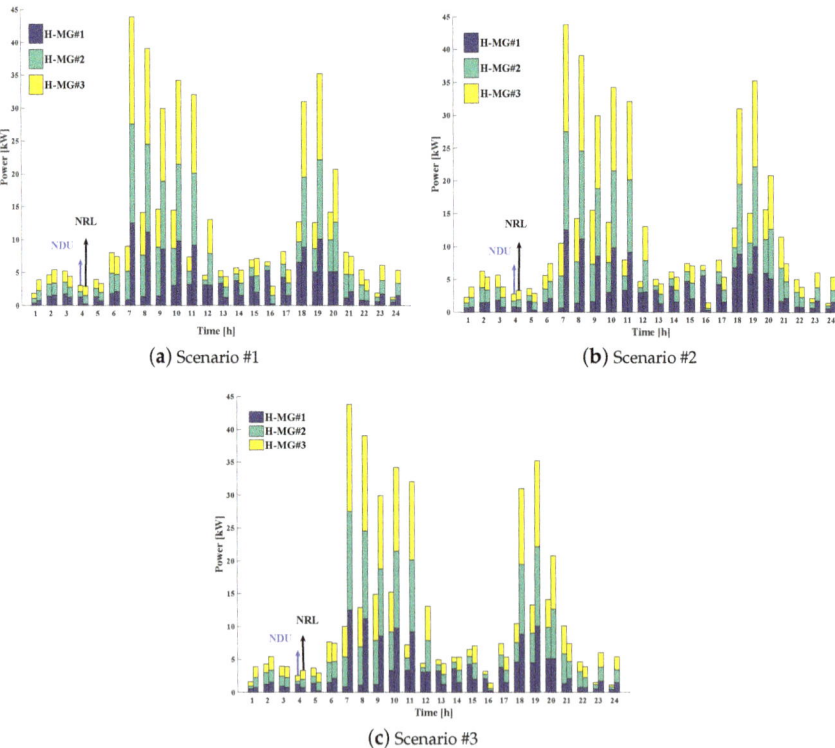

Figure 7. NDU and NRL power profiles of each H-MG.

In Figure 8, the generated power by DGUs of each H-MG is illustrated. Despite the higher load in H-MG #2 and #3 compared to H-MG #1, the amount of DGU's generation by H-MG #1 is much higher than other H-MGs during the early hours of the morning. As can be seen in Figure 7, in these hours, the amount of generated power by NDUs of H-MG #1 is much less than other H-MGs. Therefore, the shortage of H-MG #2 and H-MG #3 is supplied through DGU of H-MG #1. A comparison of the results of DGUs in Scenarios #2 and #3 indicates that according to the increase in the generated power from renewable resources in Scenario #2, the DGUs' production capacity in this scenario should be less than scenario #3. However, in a few time intervals, the algorithm has decided that the amount of generated power by DGUs in Scenario #2 could be higher than its amount in Scenario #3. This difference is very noticeable at 10 AM. In addition, owing to the fact that the amount of RLD has increased by 71% in Scenario #2, ES of H-MG #1 in Scenario #2 has discharged twice as much as Scenario #3. Indeed, the algorithm has striven to feed it. For the rest of the day, there is no noticeable change in the amount of generated power by DGUs in all H-MGs.

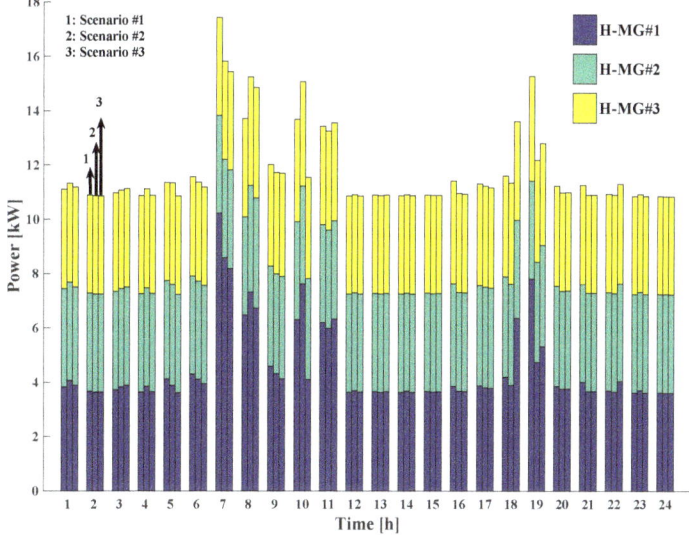

Figure 8. DGU power profile of each H-MG.

The power of ES in charging/discharging mode during 24-h system operation is shown in Figure 9. At some intervals, due to the sudden decline in the power generation from renewable resources, the algorithm has preferred to use the ES in order to meet the demand of H-MGs. On the other hand, if there is excess power in the system, this surplus power usually is used by the algorithm to charge the ESs in the network in order to maintain the state of charge (SOC) of the batteries at their maximum values. This approach will significantly boost the reliability of the system in response to power shortages or encountered unwanted events at other times. Based on this strategy, all ESs in the system will be set at their maximum value for their operation in the next day.

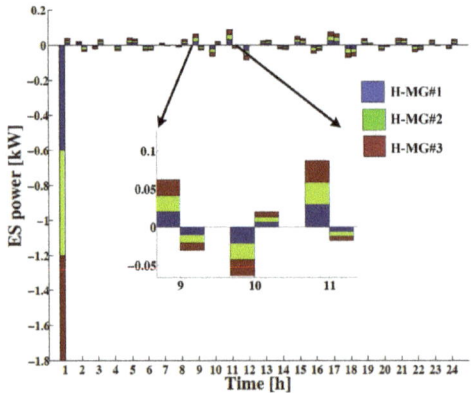

Figure 9. ES power of each H-MG in charging/discharging mode.

One of the main advantages of the proposed algorithm is its ability to control the RLD. The amount of RLD at different time intervals of a day is shown in Figure 10. As can be seen, at the early hours of the morning, when the NRL is very high, the algorithm has almost used the produced power by DGUs (Figure 8) and also the purchasing power from the upstream grid (as shown in Figure 11) to cover the NRL. Hence, the algorithm has allocated a small amount of power to feed the RLD.

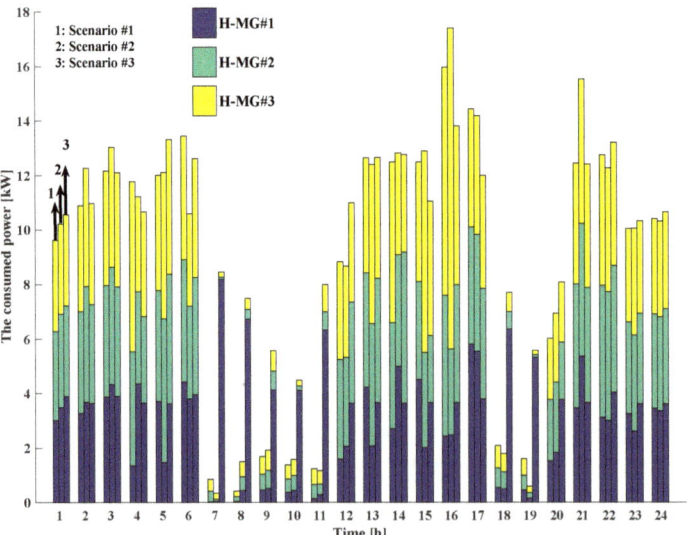

Figure 10. RLD power profile of each H-MG.

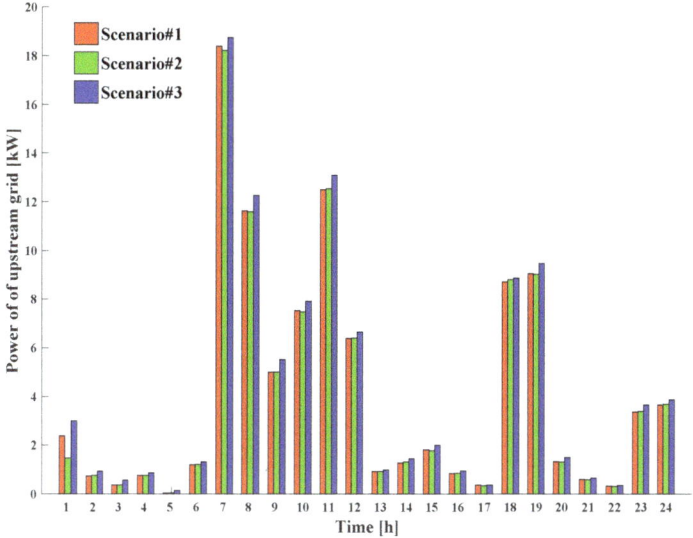

Figure 11. Upstream grid's power profile for selling.

Figure 12 shows values of the converged pay-off function for the consuming players, generating players and distribution companies under the implemented scenarios. As observed from Figure 12a, during the time interval of 7:00–8:00 am, all H-MGs experience power shortage and accept a cost for compensating the value of power demand from the upstream grid. As a result, they cannot gain revenue by selling power to their consumers and/or other H-MGs as observed in Figure 12b.

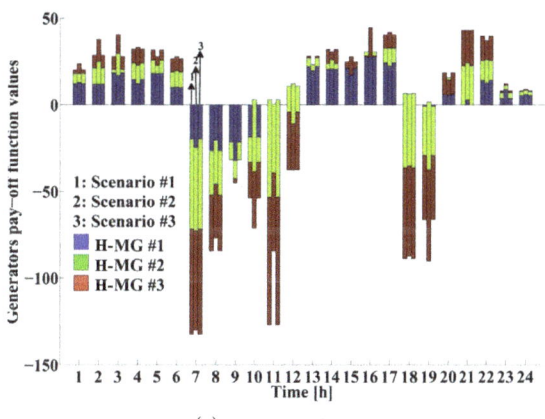

(**a**) Generating players

Figure 12. *Cont.*

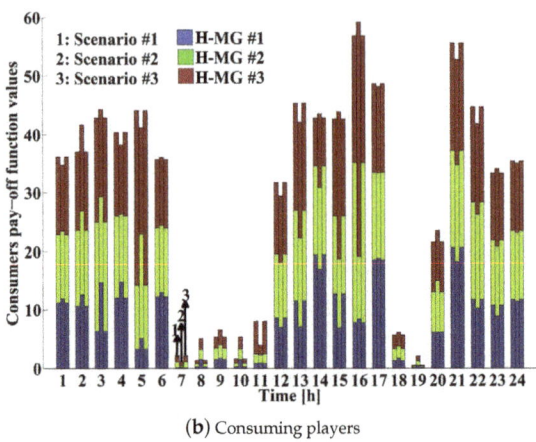

(b) Consuming players

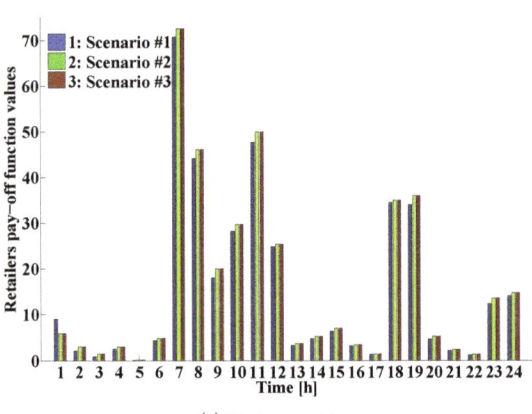

(c) Upstream grid

Figure 12. Pay-off function vs. time.

However, the amount of revenue of the upstream grid has increased significantly during this time interval as observed in Figure 12c. Also, during some time intervals, some of the H-MGs are observed to gain revenue but other H-MGs are charged for supplying their load demand. During these time intervals, H-MGs having excess power gain revenue by selling the required power to the H-MG encountering a power shortage. Furthermore, the upstream grid also compensates for the remaining power required by H-MGs having a power shortage. Thus, the revenue resulting from selling electricity is obtained.

In Scenario #1 and #2, H-MG #1 gained profit by selling power during 87.5% of the time intervals in a day. However, just during 25% of this time period, its revenue has been obtained from other H-MGs. This is why during this scenario, H-MG #2 gained profit from 62.5% of the time period. This value has reached about 54% for H-MG #3. With the reduction of power generated by renewable resources (in Scenario #3), the amount of H-MG #1 revenue has decreased by about 10%. This reduction in H-MGs #2 and #3 is about 7%. As it is observed from Figure 12b since the payoff function related to the consuming players is based on the reduction of electricity cost (during time intervals which the algorithm has increased the value of RLD demand for all H-MGs), the payoff function of the consumers has also increased.

To evaluate the performance and capability of the proposed algorithm in improving H-MGs pay-off in MH-MGs, its hourly value in the single H-MG system connected to the upstream grid and in the MH-MG network, shown in Figure 13, is evaluated. For this reason, values of the payoff function of H-MG #1 investigated in two case studies (single H-MG and MH-MG network) are evaluated. Although in the range of some intervals, the value of the pay-off in the single H-MG network is more than or equal to its value in the MH-MG network; however, a 78% increase in its value is observed in MH-MG during the 24 h period. It is imperative to state this point because the cost accepted by H-MG #1 during the time intervals for buying power is much less than its value in the single H-MG network.

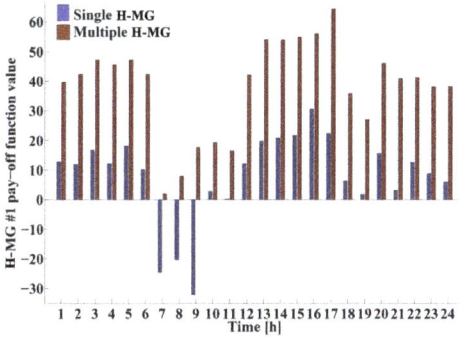

Figure 13. Pay-off function related to H-MG #1 in the single H-MG and MH-MG under Scenario #1.

In addition, to assess the effectiveness of the proposed algorithm, an independent simulation test in comparison to the ICA [18] under the normal operation has been conducted. The total payoff of all players under uncertainties in the network that consists of two H-MGs connected to the upstream grid is reported in Table 1. As the numeric results demonstrate that the REM-NIRA has been successful to achieve approximately a 169% boost in the total payoff related to the ICA. This outcome asserts that the REM-NIRA is able to improve the performance of the market with different ownership and contradictory objectives as well as power distribution in the network. Hence, more stakeholders are persuaded to engage in energy trading and as a consequence, the competition would increase significantly. Furthermore, this structure can assist in reducing electricity cost.

Table 1. Total payoff values of all players related to REM-NIRA and imperialist competition algorithm (ICA) under Scenario #1.

Objective	REM-NIRA	ICA
Total payoff value	18.52	6.89

8. Conclusions

A centralized economic structure was proposed for MH-MG systems in this study. The proposed structure connected to the upstream grid was evaluated considering different objective functions including generating and consuming players separately. For each H-MG, the proposed structure provided an optimum scheduling for exchanging power among H-MGs while satisfying the defined objective functions and technical constraints. Presenting a fair non-cooperative structure like this, encourages a wide range of players with different ownership to take part actively in a competition of energy trading that could form the basis for creating an interactive and a powerful structure in the future power networks.

The discussed problem was formulated as a general multi-objective optimization problem and an algorithm based on the NIRA method was presented for solving the problem searching for a way to understand the electricity buyers and sellers' individual behaviors and discover the optimal strategies which lead to maximizing the pay-off of all these players with contradicting goals in the competitive market. Interestingly, the formulated problem has very simplified formulas with smaller problems and less computational complexities relative to its dimensions. The proposed algorithm has the capability to exchange the optimum power in the H-MG distribution system where power management and extra load sharing functions were at no extra cost. The proposed algorithm increased the H-MGs' interaction with one another and with the upstream grid by increasing the profit, reducing power mismatch, and reducing the electricity market clearing price. It was argued that the proposed structure can easily be applied to other scenarios with alternative aims and constraints rather than cases discussed in this paper.

The obtained numerical results showed that the presented structure will result in the minimum cost and consequently the maximum profit for players during their performance as consuming and generating players. Moreover, various flexibility resources and numerous players can be accommodated conveniently in order to address the concept of maintaining equilibrium state of a system between the local power supply and load demand, ergo, the proposed algorithm could offer technical advantages for a real-time power management of H-MGs to assure safe exploitation, distribution optimization and demand side management. Additionally, it could be used as an assured and effective programming tool for managing risk and investment studies since it could estimate the power dispatch profile of the generating resources which are either dependent or independent of loads, stochastic power and renewable resources.

In future research, authors are going to make advances on the REM-NIRA performance by providing cooperation opportunities between diverse partakers to join coalitions in the market through a dynamic binding strategy. Furthermore, the optimal power flow restrictions like voltage at different locations and also carbon emission constraints will be considered in the mathematical model.

Author Contributions: All authors jointly contributed to the research model and implementation, results analysis and writing of the paper.

Funding: This research received no external funding.

Conflicts of Interest: The authors declare no conflict of interest.

Nomenclature

	Acronyms
CEMS	central energy management system
DER	distributed energy resources
DGU	dispatchable generation unit
DNO	distribution network operator
EMS	energy management system
ES	energy storage
ES+, ES−	ES during charging/discharging mode
EV	expected value
GR	upstream grid
H-MG	home microgrid
H-MG+, H-MG−	surplus/shortage power of H-MG
ICA	imperialist competition algorithm
MCEMS	modified conventional energy management system
MCP	market clearing price
MH-MG	multiple home microgrid
MO	market operator
MT	micro-turbine
NDU	non-dispatchable unit
NRL	non-responsive load
PBUC	price-based unit commitment
PV	Photo-voltaic

SOC	state of charge
REM	retail energy market
REM-NIRA	REM based on Nikaido-Isoda/relaxation algorithm
RLD	responsive load demand
TOAT	Taguchi's orthogonal array testing
WT	Wind turbine

Sets and Indices

θ, β	load demand curve coefficients
a^j, b^j, c^j	coefficients of cost function of DGU in H-MG #j
q, q', q''	number of generating/consuming/distribution companies players
N_s	the number of the uncertainty
s	the scenario of A
n	number of H-MGs
π^{ES-}, π^{ES+}	the supply bids by ES−/ES+ ($/kWh)
Δt	time interval

Constants

$\overline{P}^{A,j}, \underline{P}^{A,j}$	the maximum /minimum output power of A in H-MG #j (kW)
	A ∈ {ES−, ES+, DGU, NDU, H-MG−, H-MG+, NRL, RLD}
$\overline{SOC}^{ES,j}, \underline{SOC}^{ES,j}$	limit of SOC of ES in H-MG #j (%)

Parameters

$\lambda^{GR,i''}(t)$	offer price of distribution grid #i'' at time t ($/kWh)
$P_s^{A,j}(t)$	output power of resource A under scenario #s in the H-MG #j (kW)
$\rho_s^{A,j}(t)$	probability of scenario #s of resource A in the H-MG #j

Functions

$C^i(t), R^i(t), J^i(t)$	cost/revenue/profit functions of generating player #i at time t ($) (i∈ {1,2,⋯,q})
$C^{A,j}(t)$	cost of producing/buying power in H-MG #j ($)
$C^{GR,i''}(t), R^{GR,i''}(t), J^{GR,i''}(t)$	cost/revenue/profit functions of distribution grid #i'' ($) (i∈ {1,2,⋯,q})
$C^{H-MG+,jm}(t)\|_{m \ne j},$	cost of buying power by H-MG #j from H-MG #m/distribution grid #i'' ($)
$C^{H-MG+,ji''}(t)$	
	(i'' ∈ {1,2,⋯,q''})
$J^{i'}(t)$	profit functions of consuming player #i' at time t ($)
$\lambda^{H-MG,j}(t)$	offer price of H-MG #j at time t ($/kWh)
$EV^{A,j}(t)$	expected value of A in H-MG #j at time t
ΔP	amount of shortage power of H-MG #j is supplied partly or totally by the excess power of H-MG #m

Decision Variables

$p^{A,j}(t)$	output power of A in H-MG #j during the time period t (kWh)
$X^{H-MG+,j}(t)$	decision making variable of H-MG #j (i.e., 0 if H-MG #j is not satisfied through power exchange with other H-MGs and 1 if otherwise)
$p^{H-MG+,ji''}(t), p^{H-MG-,ji''}(t)$	amount of power which distribution grid #i'' sells /buys to/from H-MG #j at time t (kW)
x*	Nash equilibrium
$SOC^{ES,j}(t)$	ES SOC of H-MG #j at time t (%)

Appendix A

The details of the test system are presented in Table A1. Also, Table A2 provides the features of the devices of every H-MG and the coefficients related to the load demand prices.

Table A1. The input data of the proposed game structure.

Input Data	Value in the Test System
Number of H-MGs	3
Number of players	7
Type of game	static
Players' dimensions vector	[4, 1, 4, 1, 4, 1, 2]
Upper bound level of players	∞
Lower bound level of players	0
Termination tolerance	1×10^{-5}
Maximum number of iterations allowed by the relaxation algorithm	100

Table A2. Rated profile of distributed energy resources (DERs).

Parameter	Value	Symbol
ES System		
Maximum ES power during dis/charging modes (kW)	$\overline{P}^{ES+}/\overline{P}^{ES-}$	0.816/3.816
Initial state of charge (SOC) at T (%)	SOC_I	50
Maximum/minimum SOC (%)	$\overline{SOC}/\underline{SOC}$	80/20
Initial stored energy in ES (kWh)	E_I^{ES}	1
Total capacity of ES (kWh)	E_{Tot}^{ES}	2
Consumer bid by ES+ ($/kWh)	π_t^{ES+}	0.145
Photo-Voltaic (PV)		
Maximum/minimum instantaneous power for PV (kW)	$\overline{P}^{PV}/\underline{P}^{PV}$	6/0
Wind Turbine (WT)		
Maximum/minimum instantaneous power for WT (kW)	$\overline{P}^{WT}/\underline{P}^{WT}$	8/0.45
Micro-Turbine (MT)		
Maximum/minimum instantaneous power for MT (kW)	$\overline{P}^{MT}/\underline{P}^{MT}$	12/3.6
Coefficients of cost function of DGU	a($/kW^{2}h)	[$6 \times 10^{-6}, 7 \times 10^{-6}, 8 \times 10^{-6}$]
	b($/kWh)	[0.01, 0.015, 0.013]
	c($/h)	0
Load Coefficients		
Load demand curve coefficients	θ	0.001
	β	3.4
Maximum coefficient of RLD related to NRL	ζ	15

References

1. Bashir, A.A.; Pourakbari Kasmaei, M.; Safdarian, A.; Lehtonen, M. Matching of Local Load with On-Site PV Production in a Grid-Connected Residential Building. *Energies* **2018**, *11*, 2409. [CrossRef]
2. Al-Sumaiti, A.S.; Salama, M.M.; El-Moursi, M. Enabling electricity access in developing countries: A probabilistic weather driven house based approach. *Appl. Energy* **2017**, *191*, 531–548. [CrossRef]
3. Marzband, M.; Yousefnejad, E.; Sumper, A.; Domínguez-García, J.L. Real time experimental implementation of optimum energy management system in standalone Microgrid by using multi-layer ant colony optimization. *Int. J. Electr. Power Energy Syst.* **2016**, *75*, 265–274. [CrossRef]
4. Shafiee, Q.; Dragičević, T.; Vasquez, J.C.; Guerrero, J.M. Hierarchical Control for Multiple DC-Microgrids Clusters. *IEEE Trans. Energy Convers.* **2014**, *29*, 922–933. [CrossRef]
5. Lu, X.; Guerrero, J.M.; Sun, K.; Vasquez, J.C.; Teodorescu, R.; Huang, L. Hierarchical Control of Parallel AC-DC Converter Interfaces for Hybrid Microgrids. *IEEE Trans. Smart Grid* **2014**, *5*, 683–692. [CrossRef]
6. Marzband, M.; Azarinejadian, F.; Savaghebi, M.; Guerrero, J.M. An Optimal Energy Management System for Islanded Microgrids Based on Multiperiod Artificial Bee Colony Combined With Markov Chain. *IEEE Syst. J.* **2017**, *11*, 1712–1722. [CrossRef]
7. Ou, T.C.; Hong, C.M. Dynamic operation and control of microgrid hybrid power systems. *Energy* **2014**, *66*, 314–323. [CrossRef]
8. Acharya, S.; Moursi, M.S.E.; Al-Hinai, A.; Al-Sumaiti, A.S.; Zeineldin, H. A Control Strategy for Voltage Unbalance Mitigation in an Islanded Microgrid Considering Demand Side Management Capability. *IEEE Trans. Smart Grid* **2018**. [CrossRef]
9. Ou, T.C.; Lu, K.H.; Huang, C.J. Improvement of Transient Stability in a Hybrid Power Multi-System Using a Designed NIDC (Novel Intelligent Damping Controller). *Energies* **2017**, *10*, 488. [CrossRef]
10. Ou, T.C. A novel unsymmetrical faults analysis for microgrid distribution systems. *Int. J. Electr. Power Energy Syst.* **2012**, *43*, 1017–1024. [CrossRef]
11. Ou, T.C. Ground fault current analysis with a direct building algorithm for microgrid distribution. *Int. J. Electr. Power Energy Syst.* **2013**, *53*, 867–875. [CrossRef]

12. Koolen, D.; Sadat-Razavi, N.; Ketter, W. Machine Learning for Identifying Demand Patterns of Home Energy Management Systems with Dynamic Electricity Pricing. *Appl. Sci.* **2017**, *7*, 1160. [CrossRef]
13. Ahmad, A.; Khan, A.; Javaid, N.; Hussain, H.M.; Abdul, W.; Almogren, A.; Alamri, A.; Azim Niaz, I. An Optimized Home Energy Management System with Integrated Renewable Energy and Storage Resources. *Energies* **2017**, *10*, 549. [CrossRef]
14. Yener, B.; Taşcıkaraoğlu, A.; Erdinç, O.; Baysal, M.; Catalão, J.P.S. Design and Implementation of an Interactive Interface for Demand Response and Home Energy Management Applications. *Appl. Sci.* **2017**, *7*, 641. [CrossRef]
15. Asaleye, D.A.; Breen, M.; Murphy, M.D. A Decision Support Tool for Building Integrated Renewable Energy Microgrids Connected to a Smart Grid. *Energies* **2017**, *10*, 1765. [CrossRef]
16. Hussain, H.M.; Javaid, N.; Iqbal, S.; Hasan, Q.U.; Aurangzeb, K.; Alhussein, M. An Efficient Demand Side Management System with a New Optimized Home Energy Management Controller in Smart Grid. *Energies* **2018**, *11*, 190. [CrossRef]
17. Tushar, W.; Yuen, C.; Mohsenian-Rad, H.; Saha, T.; Poor, H.V.; Wood, K.L. Transforming Energy Networks via Peer-to-Peer Energy Trading: The Potential of Game-Theoretic Approaches. *IEEE Signal Process. Mag.* **2018**, *35*, 90–111. [CrossRef]
18. Marzband, M.; Parhizi, N.; Savaghebi, M.; Guerrero, J. Distributed Smart Decision-Making for a Multimicrogrid System Based on a Hierarchical Interactive Architecture. *IEEE Trans. Energy Convers.* **2016**, *31*, 637–648. [CrossRef]
19. Arun, S.L.; Selvan, M.P. Intelligent Residential Energy Management System for Dynamic Demand Response in Smart Buildings. *IEEE Syst. J.* **2018**, *12*, 1329–1340. [CrossRef]
20. Wu, X.; Hu, X.; Yin, X.; Moura, S.J. Stochastic Optimal Energy Management of Smart Home With PEV Energy Storage. *IEEE Trans. Smart Grid* **2018**, *9*, 2065–2075. [CrossRef]
21. Jia, L.; Tong, L. Dynamic Pricing and Distributed Energy Management for Demand Response. *IEEE Trans. Smart Grid* **2016**, *7*, 1128–1136. [CrossRef]
22. Wei, W.; Liu, F.; Mei, S. Energy Pricing and Dispatch for Smart Grid Retailers Under Demand Response and Market Price Uncertainty. *IEEE Trans. Smart Grid* **2015**, *6*, 1364–1374. [CrossRef]
23. Tavakoli, M.; Shokridehaki, F.; Akorede, M.F.; Marzband, M.; Vechiu, I.; Pouresmaeil, E. CVaR-based energy management scheme for optimal resilience and operational cost in commercial building microgrids. *Int. J. Electr. Power Energy Syst.* **2018**, *100*, 1–9. [CrossRef]
24. Nunna, H.S.V.S.K.; Doolla, S. Demand Response in Smart Distribution System With Multiple Microgrids. *IEEE Trans. Smart Grid* **2012**, *3*, 1641–1649. [CrossRef]
25. Eksin, C.; Deliç, H.; Ribeiro, A. Demand Response Management in Smart Grids With Heterogeneous Consumer Preferences. *IEEE Trans. Smart Grid* **2015**, *6*, 3082–3094. [CrossRef]
26. Melgar Dominguez, O.D.; Pourakbari Kasmaei, M.; Lavorato, M.; Mantovani, J.R.S. Optimal siting and sizing of renewable energy sources, storage devices, and reactive support devices to obtain a sustainable electrical distribution systems. *Energy Syst.* **2018**, *9*, 529–550. [CrossRef]
27. Marzband, M.; Ghazimirsaeid, S.S.; Uppal, H.; Fernando, T. A real-time evaluation of energy management systems for smart hybrid home Microgrids. *Electr. Power Syst. Res.* **2017**, *143*, 624–633. [CrossRef]
28. Valinejad, J.; Marzband, M.; Akorede, M.F.; Barforoshi, T.; Jovanović, M. Generation expansion planning in electricity market considering uncertainty in load demand and presence of strategic GENCOs. *Electr. Power Syst. Res.* **2017**, *152*, 92–104. [CrossRef]
29. Marzband, M.; Azarinejadian, F.; Savaghebi, M.; Pouresmaeil, E.; Guerrero, J.M.; Lightbody, G. Smart transactive energy framework in grid-connected multiple home microgrids under independent and coalition operations. *Renew. Energy* **2018**, *126*, 95–106. [CrossRef]
30. Marzband, M.; Fouladfar, M.H.; Akorede, M.F.; Lightbody, G.; Pouresmaeil, E. Framework for smart transactive energy in home-microgrids considering coalition formation and demand side management. *Sustain. Cities Soc.* **2018**, *40*, 136–154. [CrossRef]

31. Abdelsalam, A.A.; Gabbar, H.A.; Musharavati, F.; Pokharel, S. Dynamic aggregated building electricity load modeling and simulation. *Simul. Model. Pract. Theory* **2014**, *42*, 19–31. [CrossRef]
32. Marzband, M.; Javadi, M.; Domínguez-García, J.L.; Mirhosseini-Moghaddam, M. Non-cooperative game theory based energy management systems for energy district in the retail market considering DER uncertainties. *IET Gener. Transm. Distrib.* **2016**, *10*, 2999–3009. [CrossRef]

© 2018 by the authors. Licensee MDPI, Basel, Switzerland. This article is an open access article distributed under the terms and conditions of the Creative Commons Attribution (CC BY) license (http://creativecommons.org/licenses/by/4.0/).

Article

Integrated Energy System Configuration Optimization for Multi-Zone Heat-Supply Network Interaction

Bo Tang [1,*], Gangfeng Gao [1], Xiangwu Xia [1,2] and Xiu Yang [1]

1. School of Electric Power Engineering, Shanghai University of Electric Power, Shanghai 200090, China; gaogangfeng@mail.shiep.edu.cn (G.G.); xiaxiangwu@mksh.com.cn (X.X.); yangxiu721102@126.com (X.Y.)
2. China Coal Technology & Engineering Group Shanghai Research Institute, Shanghai 200030, China
* Correspondence: tangbo@shiep.edu.cn; Tel.: +86-135-6461-0981

Received: 20 September 2018; Accepted: 2 November 2018; Published: 6 November 2018

Abstract: The integrated energy system effectively improves the comprehensive utilization of energy through cascade utilization and coordinated scheduling of various types of energy. Based on the independent integrated energy system, the thermal network interaction between different load characteristic regions is introduced, requiring a minimum thermal grid construction cost, CCHP investment operation cost and carbon emission tax as the comprehensive optimization targets, and making overall optimization to the configuration and operation of the multi-region integrated energy systems. This paper focuses on the planning of equipment capacity of multi-region integrated energy system based on a CCHP system and heat network. Combined with the above comprehensive target and heat network model, a mixed integer linear programming model for a multi-region CCHP system capacity collaborative optimization configuration is established. The integrated energy system, just a numerical model solved with the LINGO software, is presented. Taking a typical urban area in Shanghai as an example, the simulation results show that the integrated energy system with multi-zone heat-suply network interaction compared to the single area CCHP model improved the clean energy utilization of the system, rationally allocates equipment capacity, promotes the local consumption of distributed energy, and provides better overall system benefits.

Keywords: integrated energy system; thermal network planning; carbon emission; clean energy; energy storage device

1. Introduction

A regional integrated energy system (RIES) is a comprehensive regional energy supply network formed by the coupling of single energy systems such as electricity, gas and thermal (cold). It is a clean, economic, efficient and environmentally friendly energy supply system at the present stage [1,2]. The multi-energy complementarity, synergistic optimization and energy cascade utilization of the integrated energy system have improved the comprehensive energy efficiency and reduced the emissions of air pollutants. At the same time, the integrated energy link among multiple regions balances the difference of energy use between regions, cuts peak loads and fills valleys, and improves energy supply reliability [3,4]. However, the complex structure of comprehensive energy system, the coupling of various energy sources, and the matching of installed capacity directly affect the economy of the system operation and the adjustment strategy of unit operations, which is of great theoretical significance and application value for the allocation of multiple energy resources and the optimization of operation strategies in the comprehensive energy system [5–9].

Reference [10] comprehensively coordinated multi-energy forms on the supply side and demand side, and carried out coordinated planning by adopting the grid method to achieve multi-energy complementarity and energy cascade utilization. Reference [11] fully considered the uncertainty of renewable energy output and terminal loads in the integrated energy system, and studied the scheduling optimization of integrated energy systems based on interval linear programming. In [12,13], under different operating modes and different load structures, the optimal configuration and comprehensive operation efficiency of the CCHP system combined with energy storage devices are studied. References [14,15] consider the coordinated planning and operation of the multi-area CCHP system of the heat-supply network model, which improves the gas turbine utilization rate, reduces the gas boiler configuration capacity and the thermal energy transmission loss, and significantly reduces the operating cost. In [16], the energy storage device concept is introduced into the distributed coordination system to optimize the configuration of the mixed integer linear programming model to realize the synchronization optimization of the system structure and operation, and the simultaneous optimization of each device and the energy storage device. References [17–20] combine solar energy with CCHP systems to optimize the number of units and gas turbine capacity of the optimal joint supply system under different operational control strategies, and achieves the maximum comprehensive benefit of the multi-target joint supply system. Reference [21] proposed a multi-objective optimization model for urban integrated electrical power, thermal and gas grids, which is used for control optimization of modified PRS integrated with thermal users, and thermally integrates the entire system with the district heating network, the effectiveness of the model in terms of economic and environmental performance is quantified by software. With the advancement of the carbon emission trading mechanism, references [22–24] comprehensively consider the impact of carbon emissions on the CCHP system, and establish a low-carbon scheduling multi-objective optimization model for carbon trading costs, fuel costs and environmental costs. Reference [25] integrates renewable energy (RE) into an autonomous CCHP system to simultaneously achieve zero environmental emissions and higher power generation and energy efficiency advantages, using an evolutionary particle swarm optimization algorithm to optimize the different configuration size of the autonomous RE-CCHP system. Reference [26] uses the Analytic Hierarchy Process (AHP) to optimize the configuration of hybrid CCHP systems considering three objective functions: annual operating cost ratio (AOCR), primary energy saving ratio (PESR) and carbon emission reduction rate (CERR). Coupling of cold, heat and electric power loads between multi-region CCHP systems, coordinated planning and optimized operation of multi-zone systems, higher utilization of equipment than single CCHP systems, reduced configuration capacity, and significantly reduced operating costs, enabling multiple systems "Multiple horizontal complementarity, vertical source network charge and storage coordination". However, at present, there are few studies on multi-energy networks such as inter-area multi-energy flow access to grids, heat networks, gas networks, etc., in terms of mixed power flow and optimal scheduling among different energy networks [27–30].

The above references mostly optimize the operation scheduling of integrated energy systems under the condition of known equipment capacity, and research on the collaborative planning of equipment capacity between various energy sources and between multiple regions is lacking. In this paper, a heat network is ingeniously introduced to regions of different load characteristics, In the references, the heating network construction of a single area is generally carried out. Based on the thermal energy-flow constraint, a simplified heat network model is established, which implements thermal energy coupling between different regions. The CCHP system takes into account the combination of devices with different characteristics, establishes an objective function which is considered operating cost, unit investment cost and carbon emission tax. The thermal energy interaction and pipeline flow changes of CCHP systems in various regions are analyzed with examples. The collaborative configuration optimizes regions with different load characteristics. The capacity allocation and network structure of various energy facilities are optimized separately to minimize the total energy supply cost of the system. In the aspect of configuration planning optimization, the redundancy of similar equipment between regions is reduced, and the centralized energy supply characteristics of high-efficiency equipment are reflected. The cost of inter-regional collaborative optimization operation is lower than that of the original independent system operation, the carbon emission tax is reduced, and the clean energy consumption is increased. To a certain extent, the overcapacity or shortage of individual regions is eliminated, and the complementary characteristics of supply and demand between regions are visually reflected.

2. Materials and Methods

2.1. Integrated Energy System Structure

In the integrated energy system, the three types of loads, including electricity, cold and thermal, are mainly provided by electric power grids, gas turbines (GT), and gas boilers (GB). The photovoltaic generator set and solar collector are added to the system plan to form a comprehensive solar energy utilization and supply system (PVCU CCHP). Multi-energy can be supplied to the cold, thermal and electrical load through different energy transfer mediums "Electrical, Thermal, and Cold Bus" [13]. The multi-energy flow structure of the integrated energy system is shown in Figure 1.

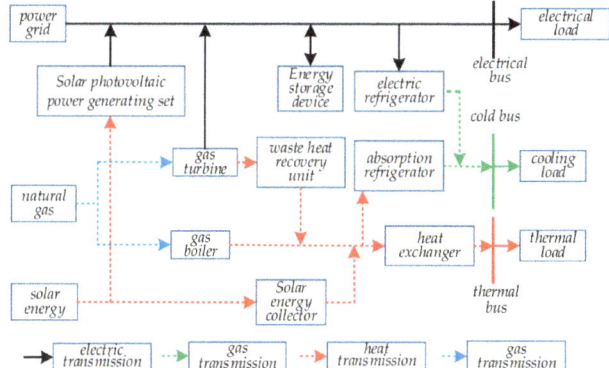

Figure 1. Integrated energy system multi-energy flow structure.

2.2. Heat-Supply Network Structure

The heat-supply network is used as the link to connect the thermal load between the CCHP systems. The heat-supply network is laid along the road and arranged in a ring according to the geographical location. When any pipe in annular network is damaged, a gate valve can separate it from other pipelines for maintenance, so it has higher safety reliability. The thermal network in this paper is a simple model, which is established by referring to the network node interactive power model. Figure 2 shows the thermal flow distribution in the heat-supply network pipe. The thermal loss existing in the thermal flow through the pipe is defined as $\Delta H_{ij,t}$. $H_{i,t}$ and $H_{j,t}$ represent the thermal energy at both ends ij of the pipe at time t.

2.3. Heat-Supply Network Model

This model is based on the basic principle of heat transfer and pipe networks, with the interacting power ($H_{ex,i,t}$) between each area and heat-supply network, the flow of heat $H_{ij,t}$, the node flow $q_{i,t}$ as the main factors. The optimization variables are based on the energy conservation law (flow balance) constraint, and the following simplified linearized hot water network model is established. In the short-distance region, assuming that the temperature field of the hot network is a steady-state field, the interaction node has no heat loss, and the heat energy loss is only related to the flow distance of the heat flow in the pipeline, and the feed water temperature T_H and the return water temperature T_L are constant. The time horizon and operating cost model applied are based on 24 h a day use.

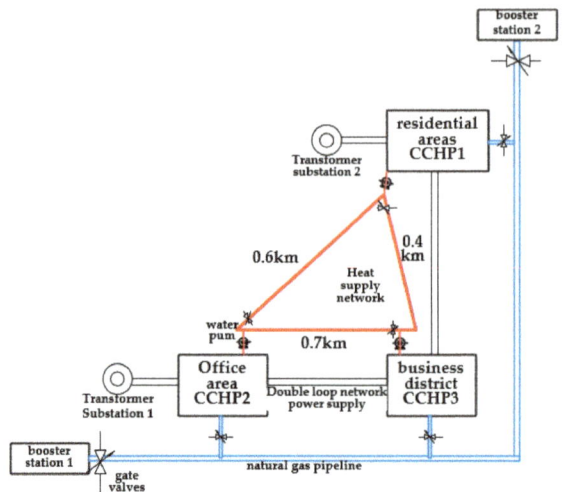

Figure 2. Area division and heat-supply network structure.

2.3.1. Basic Theory of Heat-Supply Network

(1) Thermal energy—flow constraint:

$$q_{i,t} = \frac{H_{ex,i,t}}{c(T_H - T_L)} \tag{1}$$

$$H_{ex,i,t} + \sum_{i \neq j} H_{ij,t} = 0 \tag{2}$$

where c is the specific heat capacity, taking 4187 J/(kg·°C). Equation (2) is the thermal energy balance equation of pipeline node.

(2) Thermal energy loss equation:

$$\Delta H_{ij,t} = |H_{ex,i,t} - H_{ex,j,t}| = \begin{cases} \delta L_{ij}|H_{ex,i,t}|, |H_{ex,i,t}| > |H_{ex,j,t}| \\ \delta L_{ij}|H_{ex,j,t}|, |H_{ex,i,t}| < |H_{ex,j,t}| \end{cases} \tag{3}$$

$$\sum_{i=1}^{m} H_{ex,i,t} - \sum_{\substack{i \neq j \\ i < j}} \Delta H_{ij,t} = 0 \tag{4}$$

where $\Delta H_{ij,t}$ is the heat loss of the pipeline, $H_{ex,i,t}$ and $H_{ex,j,t}$ indicate the mutual thermal energy at two ends of the pipeline at time t. δ is the thermal energy loss rate of pipeline for unit length, taking 0.1. $|H_{ex,i,t}| > |H_{ex,j,t}|$ represents thermal flow from node i to node j, and $|H_{ex,i,t}| < |H_{ex,j,t}|$ represents thermal flow from node j to node i. Equation (4) means that the sum of interactive power and pipeline thermal loss is 0.

2.3.2. Operation Cost of Heat-Supply Network Investment

(1) Investment cost of the pipeline:

$$C_{h,inv} = K_{p,fix} \sum_{i \neq j} L_{ij} + K_{p,var} \sum_{i \neq j} L_{ij} \tag{5}$$

where $K_{p,fix}$ is the cost of laying pipeline and $K_{p,var}$ is model cost of pipeline. L_{ij} is the length of pipe from i to j in heat-supply network.

(2) Electric charge for pump operation:

$$C_{h,ope} = \sum_{i=1}^{m} \sum_{t=1}^{N_t} \eta_{ehr} |H_{ex,i,t}| K_{e,i,t} \Delta t \tag{6}$$

where m is the number of CCHP co-supply systems. $K_{e,i,t}$ is the unit price of power purchase at time t for the region i. η_{ehr} is the ratio of electricity consumption to transferred thermal quantity, which means the amount of electricity consumed in the transmission of unit of thermal, taking 0.0059. Δt is the time interval, take $\Delta t = 1$ h. N_t is the number of running time periods.

2.4. CCHP Mathematical Model

The lowest total cost is just the solution of the objective function. The proposal doesn't refer to a multi-objective method. In the process of programming and solving, available energy, various types of load and the cost of unit power are input variables. The total cost of output, various types of interactive power, the rated capacity of each unit and the output power of the unit at each moment are output variables.

2.4.1. Objective Function

(1) Unit investment cost:

$$C_{cchp.inv} = \sum_{i=1}^{m} \left(\sum_{r \in \Omega_{GT}} K_{gt}^r W_{gt.i}^r x_{gt.i}^r + \sum_{r \in \Omega_{GB}} K_{gb}^r W_{gb.i}^r x_{gb.i}^r + K_{ec} W_{ec.i} x_{ec.i} + K_{ac} W_{ac.i} x_{ac.i} \right.$$
$$\left. + K_{Ppv} W_{Ppv,i} x_{Ppv.i} + K_{Hpv} W_{Hpv,i} x_{Hpv.i} + K_{ee} W_{ee,i} x_{ee.i} \right) \tag{7}$$

where K is the unit capacity price of equipment. Ω_{GT} and Ω_{GB} are GT and GB alternative model sets, respectively. $W_{gt.i}^r$, $W_{gb.i}^r$, $W_{ec.i}$, $W_{ac.i}$, $W_{Ppv,i}$, $W_{Hpv,i}$, $W_{ee,i}$ are respectively gas turbine sets, gas boilers, electric refrigerator, absorption refrigerator, photovoltaic generating sets, photovoltaic collector, electrical storage device capacity. $x_{gt.i}^r$, $x_{gb.i}^r$, $x_{ec.i}$, $x_{ac.i}$, $x_{Ppv.i}$, $x_{Hpv.i}$, $x_{ee.i}$ respectively represent the corresponding model equipment state of each equipment, taking values of 0 and 1.

(2) Operating cost:

$$C_{cchp,ope} = \sum_{i=1}^{N_e} \sum_{t=1}^{N_t} \frac{1}{H_{ng}} \left[\sum_{r \in \Omega_{GT}} \frac{P_{gt.i,t}^r}{\eta_{gt,i}^r} + \sum_{r \in \Omega_{GB}} \frac{H_{gb.i,t}^r}{\eta_{gb,i}^r} \right] \Delta t K_f + \sum_{i=1}^{N_e} \sum_{t=1}^{N_t} (K_{e,i,t} P_{ex,i,t}$$
$$+ K_{ee,i,t} |P_{ee,i,t}|) \Delta t \tag{8}$$

$$C_{om} = \sigma C_{cchp.inv} \tag{9}$$

where $C_{cchp,ope}$ is the total operating cost of the CCHP system. $K_{e,i,t}$ is the electricity purchase price of the electric grid. σ is the ratio coefficient of operation and maintenance cost, taking 0.03. C_{om} is the annual maintenance cost of the system. $P_{ex,i,t}$ is the interactive electric power between the system and the electric grid. $P_{gt.i,t}^r$ is the output power of the gas turbine, and $H_{gb.i,t}^r$ is the heat energy output of the gas boiler. $\eta_{gt,i}^r$ and $\eta_{gb,i}^r$ are the conversion efficiency of gas turbine and gas boiler, respectively. K_f is the price of natural gas, taking 2.37 Y = /m³. H_{ng} is the calorific value of natural gas, taking 9.78 (kw·h)/m³ $K_{ee,i,t}$ is the electrical storage device running loss cost, taking 0.02 Y = /(kw·h). $P_{ee,i,t}$ is the charge and discharge power of the electrical storage device.

(3) CO_2 emission tax model:

$$C_{pun} = \sum_{t=1}^{N_t}(P_{ex,i,t}\mu_e + (\sum_{r\in\Omega_{GT}}\frac{P^r_{gt,i,t}}{\eta^r_{gt,i}} + \sum_{r\in\Omega_{GB}}\frac{H^r_{gb,i,t}}{\eta^r_{gb,i}})\mu_f)C_c \quad (10)$$

where μ_e and μ_f are electricity emission factor and use gas emission factor, respectively, taking 0.8 kg/(kw·h) and 0.19 kg/(kw·h). C_c is the CO_2 emission tax, taking 0.2 RMB/kg.

This paper will combine the above optimization objectives, investment cost, operating cost and environmental cost, as the total objective function, as shown in the following expression:

$$minC = C_{cchp.inv} + C_{h,inv} + C_{h,ope} + C_{cchp,ope} + C_{om} + C_{pun} \quad (11)$$

where C is the integrated cost, and the lowest objective function value is the highest benefit. The model does not involve a separate optimization discussion of each sub-objective functions in the comprehensive target, and does not discuss the weighting of the sub-functions in the total objective function.

2.4.2. Constraints

(1) Power balance constraint.

The electric balance, thermal balance and cold balance constraints at each time slot of each load area are:

$$P_{L,i,t} = \sum_{r\in\Omega_{GT}} P^r_{gt,i,t} + P_{ex.i.t} - P_{ec,i,t} + P_{ee,i,t} + P_{pv,i,t} \quad (12)$$

$$H_{L,i,t} = \sum_{r\in\Omega_{GT}}\left(P^r_{gt,i,t}\frac{1-\eta^r_{gt,i}}{\eta^r_{gt,i}}\right)\eta_{hr,i}\eta_{he,i} + \sum_{r\in\Omega_{GB}} H^r_{gt,i,t}\eta_{he,i} - \sum H_{ac.i.t}\eta_{he,i} + H_{pv,i,t} \quad (13)$$
$$+H_{ex,i,t}$$

$$C_{L,i,t} = \sum H_{ac.i.t}E_{ac.i} + \sum P_{ec.i.t}E_{ec.i} \quad (14)$$

where $P_{L,i,t}$, $H_{L,i,t}$, $C_{L,i,t}$ are the user's demand for electricity, thermal and cold loads in area i at time t. $P_{ec,i,t}$ is the electric power consumed by electric refrigeration; $P_{pv,i,t}$ is the output power of photovoltaic generator set. $P^r_{gt,i,t}\frac{1-\eta^r_{gt,i}}{\eta^r_{gt,i}}$ is the power of residual heat by gas turbine. $\eta_{hr,i}$ is the recycling efficiency of the thermal collector, taking 0.75. $\eta_{he,i}$ is the efficiency of thermal exchanger, taking 0.9. $H_{ac.i.t}$ is the heat energy power consumed by absorption refrigeration; $H_{pv,i,t}$ is the output power of solar collector. $E_{ac.i}$ and $E_{ec.i}$ are respectively the refrigeration efficiency of electric refrigerator and absorption refrigerator for corresponding models in area i.

(2) Interactive power constraint:

$$P^{min}_{ex,i} \leq |P_{ex,i,t}| \leq P^{max}_{ex,i} \quad (15)$$

$$H^{min}_{ex,i} \leq |H_{ex,i,t}| \leq H^{max}_{ex,i} \quad (16)$$

where $P^{min}_{ex,i}$ and $P^{max}_{ex,i}$ are respectively the lower limit and upper limit for the electric power of the tie line. $H^{min}_{ex,i}$ and $H^{max}_{ex,i}$ are the lower and upper bound value of heat-supply network circulating power, respectively.

(3) Upper and lower bound for unit output:

$$\begin{cases} P^{min}_{gt,i,r} \leq P^r_{gt,i,t} \leq P^{max}_{gt,i,r} \\ H^{min}_{gb,i,r} \leq H^r_{gb,i,t} \leq H^{max}_{gb,i,r} \\ P^{min}_{ec,i} \leq P_{ec.i.t} \leq P^{max}_{ec,i} \\ H^{min}_{ac,i} \leq H_{ac.i.t} \leq H^{max}_{ac,i} \end{cases} \quad (17)$$

where $P_{gt,i,r}^{max}$ and $P_{gt,i,r}^{min}$ are the upper and lower bound of type r GT output, respectively. $H_{gb,i,r}^{max}$ and $H_{gb,i,r}^{min}$ are upper and lower bound of type r GB output, respectively. $P_{ec,i}^{max}$, $P_{ec,i}^{min}$ and $H_{ac,i}^{max}$, $H_{ac,i}^{min}$ are upper and lower bound of EC and AC output, respectively.

(4) Electrical storage device constraint:

$$S_{i,t+1} = S_{i,t} + \Delta t P_{ee,i,t}, \ t = 1, 2, \ldots, 24 \tag{18}$$

$$\begin{cases} 0.15 S_i^{man} \leq S_{i,t} \leq 0.9 S_i^{man} \\ -0.3 S_i^{man} \leq P_{ee,i,t} \leq 0.3 S_i^{man} \\ S_{i,1} = S_{i,24} \end{cases} \tag{19}$$

where $S_{i,t}$ is the storage state of the electrical storage device. $P_{ee,i,t}$ is the charging and discharging electric power at the corresponding time. In order to guarantee the service life and efficiency of the electrical storage device, the lower limit of the storage capacity of the electrical storage device is 0.15 of the total capacity, and the upper limit is 0.9 of the total capacity. Charging and discharging electric power shall not exceed 30% of the total capacity. At the same time, it is guaranteed that the storage state of the initial and ending time is the same.

(5) Photovoltaic output constraint:

$$\begin{cases} \frac{P_{pv,i,t}}{\eta_{Ppv}} + \frac{H_{pv,i,t}}{\eta_{Hpv}} \leq E_{pv,i,t} \\ 0 \leq P_{pv,i,t} \leq P_{pv,i}^{max} \\ 0 \leq H_{pv,i,t} \leq H_{pv,i}^{max} \end{cases} \tag{20}$$

where $E_{pv,i,t}$ is the available solar energy for each load area and η_{Ppv} is the efficiency of photovoltaic power generation. η_{Hpv} is the collector conversion efficiency. The upper limit of $P_{pv,i,t}$ and $H_{pv,i,t}$ are the rated power of the equipment. In conclusion, all models are integer linear models, so Lingo software is used for calculation optimization [16]. The parameters involved in this paper are shown in the Appendix A.

3. Results

This paper takes the integrated energy planning of a typical urban area in Shanghai (Caoxi area) as an example, and divides it into three load areas, including commercial area, office area and residential area according to local industrial and load characteristics, as shown in Figure 2. And all kinds of time-of-use electricity prices in Shanghai are shown in Appendix A Table A2. Caoxi region is determined by grid division of the Shanghai urban electric power network. The region is an independent electric power supply ring network unit. On this basis, thermal and cold energy supply are divided into blocks and centralized scheduling, photovoltaic power generating set and photovoltaic collector are arranged on the sunny surface of each region. Among them, the residential area has relatively high heat-to-electric ratio, which is between 1.3 and 3. The heat-to-electric ratio in the office area tends to 1 and the heat-to-electric ratio in the commercial area is around 0.5. The equipment parameters are shown in Appendix A Table A1, Various load values are shown in Appendix A Table A4.

3.1. Equipment Capacity Planning

The CCHP system without thermal network interaction is set as mode 1, and the CCHP system with thermal network interaction is set as mode 2. In the planning, considering the actual load and operation characteristics, the equipment with high utilization rate or high economic efficiency and meeting the load requirements is selected according to the optimization results, the multi-energy flow structure of the integrated energy system is redesigned. The optimized configuration and one-time investment cost contrast are shown in Table 1.

It can be seen from Table 1, after considering the heat-supply network, the configuration capacity of gas turbine of users (commercial) with low heat-to-electric ratio increases significantly, while users (residents) with high heat-to-electric ratio do not allocate gas turbine and gas boiler. When considering the heat-supply network, the users with relatively low heat-to-electric ratio can increase the output of the gas turbine to meet their own demand and reduce the electric power purchase, while the surplus thermal energy can be used to subsidize other users with relatively high heat-to-electric ratio through the heat-supply network. Therefore, the gas turbine capacity configuration in the commercial area under mode 2 increases. At the same time, for the residents with high heat-to-electric ratio in mode 1, the insufficient thermal energy can only be reignited through the gas boiler. While combined with the heat-supply network, the thermal energy can be supplied by offices and business districts, so as to reduce the output of gas boilers and reduce the configuration of gas boilers. On the other hand, the users with low heat-to-electric ratio are given priority to deploy gas turbines, and the users with high heat-to-electric ratio are given priority to deploy gas boiler. With the increase of heating network, the overall thermal power ratio of the original high thermal power ratio area will be reduced, and gas turbines will be selected for economy. Photovoltaic output and load characteristics are highly coupled, so the number of photovoltaic generating units has been greatly increased to promote the consumption of clean energy. After connecting to the heating network, the non-gas boiler configuration in residential areas increases the capacity of the photovoltaic collector to supplement the increase of daytime heat load; the gas turbine output in office and commercial areas is increased and the thermal energy supply is sufficient. Under the economic target, the two areas omit the photovoltaic collector configuration.

Table 1. Comparison of CCHP system optimization configuration and one-time investment cost.

Equipment	Residential Area		Office Area		Commercial Area		One-Time Investment Cost	
	Mode1	Mode2	Mode1	Mode2	Mode1	Mode2	Mode1	Mode2
Gas turbine/kw	1210 × 2	0	1210 × 2	1210 × 2	3515	5740	5513.5	4782.6
Gas boiler/kw	1300	0	0	0	0	0	11.3	0
Electric refrigerator/kw	50	50	0	0	0	50	4.1	8.2
Absorption refrigerator/kw	0	200	30	30	400	400	53.3	78.1
Photovoltaic generating set/kw	260	400	200	300	200	400	660	1100
Electrical storage device/kw	300	400	400	400	450	400	172.5	180
Photovoltaic collector/kw	450	650	150	0	150	0	615	533.6
Total investment cost/10,000 RMB	2413.7	1000	2118.2	2087.5	2437.8	3527.5	6969.7	6615

When planning the pipeline layout, the heat-supply network fluid flow $q_{i,j}$ is obtained using Equation (1). Then, the cross sectional area of the pipeline (to limit the fluid flow velocity in the pipeline) can be calculated from $q_{i,j} < V_{i,j} S_{i,j}$. The temperature of feed water T_H is taken as 100 °C, the backwater T_L is 70 °C, and the peak velocity is 0.6 m/s. The planning results are shown in Table 2.

By comparing the investment costs of the two modes, it can be seen that mode 2 saves capacity allocation of gas turbine and gas boiler, and mainly increases the cost of setting photovoltaic generating and laying heat-supply network pipelines. The total investment cost of mode 2 is 6859 million yuan, which is lower than that of mode 1 about 6969.7 million yuan.

Table 2. Thermal network planning.

Pipeline	Residential—Office (L_{12})	Residential—Commercial (L_{13})	Office—Commercial (L_{23})
Maximum flow $q_{i,j}$ (kg/s)	2.2 × 10^{-3}	0.011	4.4 × 10^{-3}
Pipe radius (cm)	5	9	6
Running cost of water pump (RMB)	369	109	511
Cost of pipe laying (10,000 RMB)	84.7	59.5	99.8

3.2. Operation Optimization

Through the above optimized configuration, the capacity of each regional equipment in different cases is determined. Based on the previous research, the operating output of each unit is analyzed as follows.

(1) Thermal network interaction energy and flow direction

As shown in Figure 3, after adding the heat-supply network, the commercial area with low heat-to-electric ratio generates a large amount of excess thermal energy to supply thermal to the heat-supply network while ensuring enough power supply for local area. The resident users with higher heat-to-electric ratio absorb thermal energy from the heat-supply network. The office area is close to the gas turbine heat-to-electric ratio at 9:00–13:00, and the thermal energy surplus is less and there is no thermal load during the 20:00–8:00 period. The gas turbine keeps running due to the electrical load, and supplies thermal to the heat-supply network. At 14:00–17:00, the office area absorbs thermal from the heat-supply network. The interaction between CCHP system and the heat supply network accurately reflects the configuration and operation of the energy supply equipment.

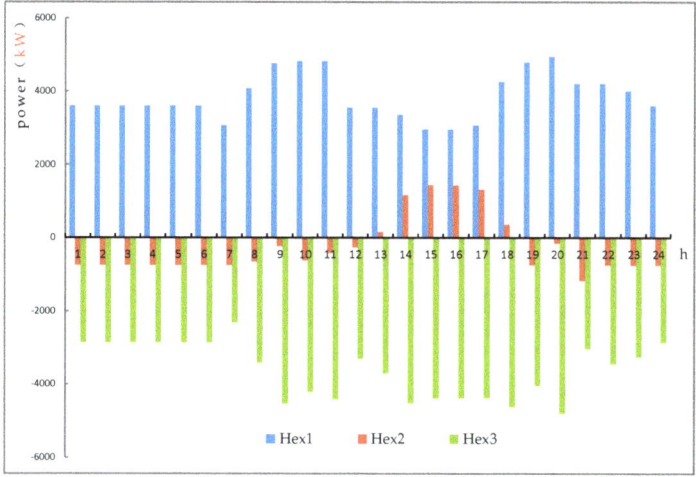

Figure 3. Interaction diagram between each CCHP system and thermal network thermal energy.

It is stipulated that the flow direction of commercial and office districts to residential areas should be positive, and the flow direction of commercial districts to office areas should be positive. In the heat-supply network planning, about 30% allowance of pipeline radius is left. It can be seen from Figure 4a that the flow direction is positive in each time period, that is, the pipeline L_{13} flows from commercial area to residential area in 24 h. The flow rate of the pipeline L_{12} is 0 during the period 13:00–18:00 with no thermal flow, and the other times are positive. The thermal energy is transported to residential areas from the office area during this period. The pipeline L_{23} is only positive at time period 13:00–18:00, and the thermal energy flows from commercial area to office area, and there is no thermal interaction at other time. Combined with the energy flow interaction analysis at each time period in Figure 3, the thermal transfer capacity of commercial and residential pipelines is much higher than that of other pipelines. However, by planning the radius of thermal network pipelines, the velocity of each pipeline is maintained at a reasonable value. This laying method reduces the inert effects and thermal loss of thermal transfer, while improving the efficiency of pipeline utilization and investment economy.

It can be seen from above that the utilization rate of commercial-office heat-supply network pipe L_{23} is low, and the pipe laying is far away and the cost is high. Therefore, considering that only the commercial-residential-office pipe should be laid, the pipeline velocity is shown in Figure 4b. After omitting the pipe L_{23}, the radius of pipe L_{23} increases from 5 to 6, the radius of pipe L_{12} remains unchanged, the pipe utilization rate increases, and the laying cost decreases.

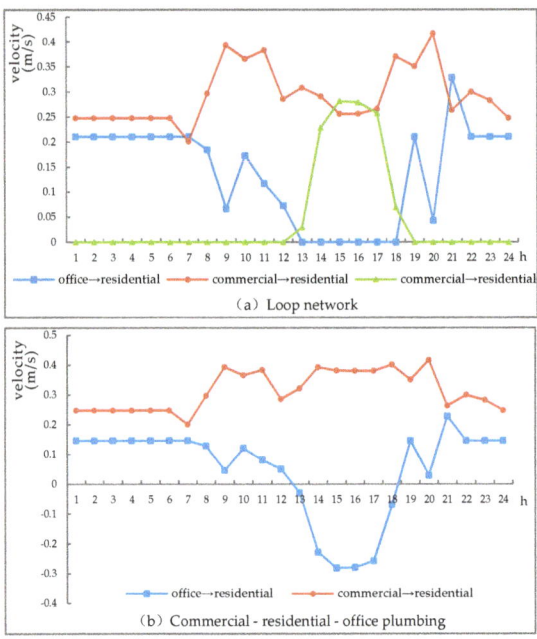

Figure 4. Flow chart of heat-supply network pipelines between regions. (**a**) Flow chart of heat-supply network pipelines of office → residential, commercial → residential, commercial → office; (**b**) Flow chart of heat-supply network pipelines of commercial → residential → office.

(2) Comparison of interaction with electric power grid

Figure 5 shows that when the residential area is in the time period 0:00–9:00 electrical load valley section in mode 1, the gas turbine operation meets the regional electric power demand. In mode 2, the electric energy is provided by the electric power grid and photovoltaic unit. During the period of 12:00–15:00, due to the increase in PV configuration capacity and the increase in photovoltaic power generation, the purchase of electricity is slightly reduced.

While the office area with relatively low heat-to-electric ratio is supplied to the thermal energy of the region, except for the period 14:00–17:00 office area gas turbine decline in output, it has increased the purchase of electricity to the electric grid. During the period from 23:00–8:00, the residential area has less electricity and high thermal load. While the commercial area has a certain electrical load and the thermal load tends to zero. At this time, the gas turbine output increases and complements the energy characteristics of the residential area. Therefore, the utilization rate of the gas turbine is increased, and it is more economical to use the distributed power source preferentially, and the difference in energy consumption is supplemented by the electric power grid.

(3) Electric power output distribution

In Figure 6a, residential area system power load is mainly supplied by the electric grid, and makes full use of photovoltaic power generation during the period 8:00–16:00, while the electricity price in the valley section 21:00–7:00, 12:00–17: 00 charges the electrical storage device. During the peak period of 8:00–11:00 and 18:00–20:00, the absorption refrigeration unit replaces the electric refrigeration unit, and the residential area absorbs more thermal from the heat-supply network. At the same time, the charged electrical storage device discharges during this period to reduce the purchase of electricity from the electric power grid and effectively reduce the peak-filling effect.

In Figure 6b, the difference between day and night electrical load of the office system is obvious. The output of the gas turbine is almost the same as that of the electric power grid. The waste thermal is recovered and supplied to the thermal load or the thermal load is supplied to the residential area through the heat-supply network. Only absorption refrigeration units and no electric refrigeration units are installed in the planning, and the output of absorption refrigeration units is maintained at a higher level during the operation. We fully utilize photovoltaic power generation from 7:00–17:00, and charge the electrical storage device at 23:00–5:00 and 13:00–18:00. During the period 8:00–12:00 and 19:00–21:00, the electrical storage device is discharged during the peak period of electricity price.

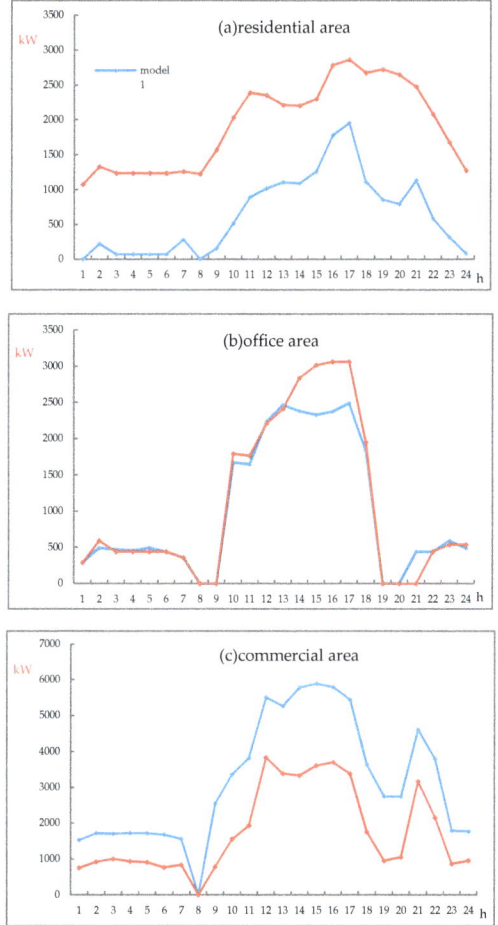

Figure 5. Electric interaction with electric power grid. (**a**) Electric interaction with electric power grid of residential area; (**b**) Electric interaction with electric power grid of office area; (**c**) Electric interaction with electric power grid of commercial area.

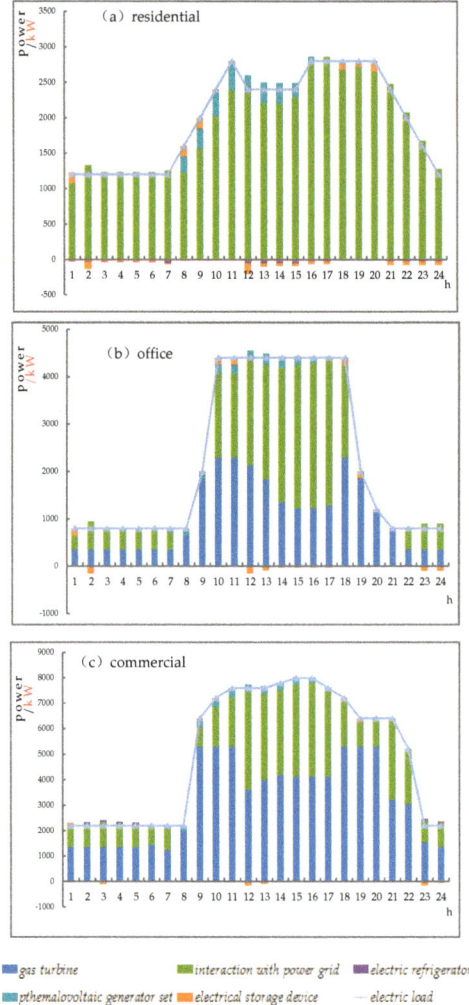

Figure 6. CCHP power output structure diagram. (**a**) CCHP power output structure diagram of residential area; (**b**) CCHP power output structure diagram of office area; (**c**) CCHP power output structure diagram of commercial area.

Similarly, in the commercial area with high heat-to-electric ratio in Figure 6c, the system electric power load is mainly satisfied by the purchase of electricity from the gas turbine and electric power grid. The gas turbine assumes the main power supply task. The electrical storage device charge and discharge and photovoltaic power generation operation conditions in this area are similar to the above two regions. The refrigeration unit operates at 23:00–5:00 during the electricity price period of the valley section, and the cooling load is independently satisfied by the absorption refrigeration unit during other periods. The load characteristics of residential and commercial areas and the analysis of the operating conditions of the equipment after planning show that the load characteristics of the two areas have strong complementary correlation characteristics. Compared with the operation of the original independent CCHP system, the joint regional planning for joining the heat-supply network reduces the repeatability of the equipment in each area while enhancing the multi-energy complementarity. The division of energy between the areas with load characteristics is clearer and reflects the operation characteristics, including the centralized energy supply, large-scale equipment and high utilization.

(4) Sensitivity of natural gas price

Figure 7 shows that as the price of natural gas rising, the operating costs of various regions also increases. The operating costs of commercial areas with increased gas turbine capacity after adding heat-supply networks increased obviously. In the later period of rising natural gas prices, the cost-saving curve is flat and the savings are decreasing which means the increase in natural gas prices has weakened the advantages of the heat-supply network. Due to the lack of electric heat-supply equipment in the system, even if the price of natural gas rises, the gas turbine must operate to maintain the energy demand of the thermal load. Therefore, when the gas turbines of each system maintain low-line operation, the utilization efficiency of the equipment is low, and the effect of increasing the comprehensive benefit of the heat-supply network to the system is limited.

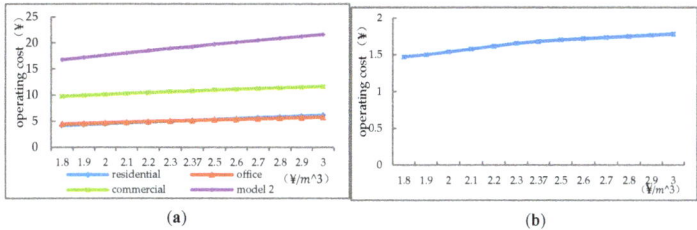

Figure 7. Operating costs of each CCHP system under different natural gas prices. (**a**) CCHP system operation costs of different areas; (**b**) the operation cost-saving of CCHP systems between two modes.

(5) Operating expenses

It can be seen from Table 3, in the planned residential areas, gas turbines and gas boilers were abandoned, the photovoltaic capacity allocation was increased, and the carbon emission tax was greatly reduced. In the office area, the difference in equipment capacity is small under mode 1 and mode 2, and the output of the equipment is similar. The carbon emission tax fluctuations in different modes are extremely small. After adding the thermal network, the gas turbine capacity in the commercial area increased, the output of the equipment increased greatly during operation, and reduced the electricity purchase from the electric grid. Compared with the large increase in output, the carbon emission tax increased slightly which indicates the cleanliness of the gas turbine operation. The capacity allocation of photovoltaic generator sets and photovoltaic collectors have a certain impact on the comprehensive benefits of the carbon emission tax. However, the two projects have a large investment cost and the unit output has uncertainty. After the addition of the heat-supply network, the total operating costs of the three regions were reduced by 8.2% compared with the independent operating costs and the carbon emission tax was reduced by 3.32%.

Table 3. CCHP system operating costs and carbon tax.

Categories	Mode 1 (RMB)			Mode 2 (RMB)			Reduction Proportion (%)
	Residential Area	Office Area	Commercial Area	Residential Area	Office Area	Commercial Area	
Carbon tax	8621	8483	19,183	7372	8464	19,245	3.32
Operating cost	51,672	50,999	107,974	-	193,368 (total)	-	8.2

4. Discussion

According to the operating characteristics, it can be seen that the gas equipment in the independent area maintains low-load operation, and the economic dispatching is limited. The joint planning of joining the heating network reduces the repeatability of equipment in each area, and the division of energy between the areas with load characteristics is more clear. It embodies the operating characteristics of centralized energy supply and high utilization of equipment.

The multi-region comprehensive energy system based on thermal network interaction not only saves the one-time investment cost, but also improves the overall operation efficiency while promoting photovoltaic absorption. Of course, the application of the heat network has its limitations. If the load characteristics between the regions are similar, the effect of increasing the heat network is not obvious, the heat storage device can be considered. And the distance between the regions should not be too long to reduce the heat energy transmission loss. In cold places, it can be used as a supplement to the distributed thermal energy system as a thermal power plant system, and the benefits brought by the construction of the heat network in the different load characteristics areas are particularly obvious.

This paper does not discuss the impact of surplus electric power inputted to electric grid and prospective annual load change on the comprehensive energy capacity allocation plan, and it is also simple to model the thermal network. The evaluation of the operation state of the comprehensive energy system, such as the primary energy utilization efficiency and environmental benefits, can be taken as the research direction of the next stage.

5. Conclusions

In this paper, the integrated energy system of typical urban residential, office and commercial mixed areas in the city is connected through the heat-supply network for joint planning, and compared with the original independent regional integrated energy system planning and operation, the following conclusions are drawn:

(1) When natural gas prices fluctuate greatly, it is not recommended to configure large-capacity gas equipment, and the appropriate capacity can be configured to meet the minimum load requirements. The load characteristics of residential and commercial areas and the analysis of the operating conditions of the equipment after planning show that the two areas are highly complementary. For areas with low heat-to-electric ratio, the combination of gas turbine and absorption chiller is prioritized to meet the cold and hot load. For areas with high heat-to-electric ratio, the output of gas turbine and gas boiler should be balanced to achieve optimal working coordination. Inter-regional heat energy interacts through the heat network, enabling multi-energy flow to achieve cross-regional distribution, coordinated planning of multi-regional systems, Although the cost of heat network pipe laying has increased about 2.44 million yuan, The total investment cost of mode 2 is lower than that of mode 1 about 1.1 million yuan, reduce proportion is 1.6%, especially the investment cost of gas turbine has been reduced about 7.3 million yuan, reduce proportion is 13.3%. Significantly reduced the configuration and investment of gas turbine and gas boiler, resulting in lower equipment costs.

(2) The primary investment cost of the electrical storage device is expensive, and the lower economic benefit brought by the electrical storage device in the area to adjust the output of the gas turbine. Therefore, the electrical storage device is mainly used to interact with the power grid, and the peak value is filled with the price response. The integration of energy storage and heat network into multi-regional joint planning can achieve multi-energy scheduling in time and space between regions. In the modeling process, a carbon emission model was added. The investment cost of photovoltaic generating set has been added about 4.4 million yuan, increased by 66.7%, it means the power of PV output increased and promotes renewable energy consumption. After the addition of the heat-supply network, the total operating costs of the three regions were reduced by 8.2% compared with the independent operating costs and the carbon emission tax was reduced by 3.32%. The simulation showed that the input of the heating network reduced the overall energy consumption of the system, promoted the use of clean energy, and improved the local consumption capacity of distributed energy.

Author Contributions: Supervision, X.Y.; investigation, resources, X.X.; software, data curation, writing—original draft preparation, G.G.; conceptualization, formal analysis, methodology, writing—review and editing, B.T.

Funding: Fund Project: Shanghai Municipal Science and Technology Commission Local Capacity Building Plan (16020500900).

Conflicts of Interest: The authors declare no conflicts of interest.

Appendix

Table A1. Equipment parameters.

Type	Unit Investment Cost (RMB /kW)	Single Rated Power	Conversion Efficiency
Pthemalovoltaic generator set	10,000	-	0.175
Pthemalovoltaic collector	8210	-	0.56
Electrical storage device	1500	-	90
Absorption chiller	1240	-	1.26
Electric refrigerator	820	-	4.13
1#gas turbine	7216.3	1210	0.243
2# gas turbine	5749.9	3515	0.279
3# gas turbine	5429.45	4600	0.293
4# gas turbine	5289.7	5740	0.32
1# gas boiler	86.84	1300	0.85
2# gas boiler	75.14	1950	0.86
3# gas boiler	73.26	4560	0.9
4# gas boiler	71.83	5200	0.9

Table A2. Electricity purchase price.

Area Type	Purchase Price/RMB		
	Peak Period	Flat Period	Valley Period
	8:00–11:00 18:00–20:00	6:00–7:00 12:00–17:00 21:00–22:00	23:00–5:00
Residential	0.677	0.377	0.377
Office, commercial	1.159	0.708	0.351

Table A3. Other parameters of integrated energy system.

Other Parameters	Resident	Commercial	Office
Installable area (m^2)	32,000	23,000	15,000
Average sunshine intensity ($kw/(M^2 \cdot d)$)	0.669	0.669	0.669
Average annual sunshine hours (h)	1970	1970	1970

Table A4. Cold and themal load and solar power available in each region (unit: kW).

Period of Time	Residential Areas				Office Area				Commercial Area			
	Electrical Load	Themal Load	Cold Load	Solar Energy	Electrical Load	Themal Load	Cold Load	Solar Energy	Electrical Load	Themal Load	Cold Load	Solar Energy
00:00–01:00	1180	3580	90	0	796	0	20	0	2194	0	224	0
01:00–02:00	1194	3577	89	0	794	0	20	0	2187	0	221	0
02:00–03:00	1192	3589	101	0	795	0	20	0	2201	0	229	0
03:00–04:00	1205	3549	96	0	801	0	20	0	2195	0	231	0
04:00–05:00	1210	3596	99	0	798	0	20	0	2197	0	236	0
05:00–06:00	1211	3602	108	0	806	0	20	0	2203	0	233	0
06:00–07:00	1198	3607	191	960	802	0	20	450	2194	0	397	690
07:00–08:00	1600	3615	195	1280	813	0	30	600	2211	0	401	920
08:00–09:00	1193	3994	204	1600	1994	2394	30	750	6397	1794	410	1150
09:00–10:00	2398	4197	197	1920	4367	4197	30	900	7204	2798	408	1380
10:00–11:00	2801	4198	201	2240	4389	4403	30	1050	7581	3609	410	1610
11:00–12:00	2423	4200	200	2560	4394	4195	30	1200	7597	3960	407	1840
12:00–13:00	2418	4204	205	2800	4402	4011	30	1350	7611	4397	417	2070
13:00–14:00	2391	3994	198	2800	4396	4004	30	1350	7799	4001	410	2070
14:00–15:00	2404	3611	192	2240	4387	3994	30	1050	8001	4013	410	1610
15:00–16:00	2790	3603	103	1600	4405	4005	30	750	7994	3998	411	1150
16:00–17:00	2797	3598	105	960	4394	4001	30	450	7598	4005	407	690
17:00–18:00	2805	3594	97	0	4385	3997	30	0	7203	4193	405	0
18:00–19:00	2800	4003	100	0	2011	2003	20	0	6414	4197	230	0
19:00–20:00	2790	4200	94	0	1206	1000	20	0	6399	4005	224	0
20:00–21:00	2414	4189	96	0	8017	500	20	0	6387	3607	223	0
21:00–22:00	2006	4176	99	0	806	0	20	0	5180	2809	219	0
22:00–23:00	1600	4000	100	0	800	0	20	0	2200	0	230	0
23:00–24:00	1200	3600	100	0	800	0	20	0	2200	0	230	0

References

1. Bie, C.; Wang, X.; Hu, Y. Review and prospect of energy Internet planning. *Proc. CSEE* **2017**, *37*, 6445–6462.
2. Yin, S.; Ai, Q.; Zeng, S.; Wu, Q.; Hao, R.; Jiang, D. Challenges and Prospects of Energy Internet Multi-energy Distributed Optimization Research. *Power Syst. Technol.* **2018**, *42*, 1359–1369.
3. Wang, W.; Wang, D.; Jia, H.; Chen, Z.; Guo, B.; Zhou, H.; Fan, M. A Summary of Steady-State Analysis of Typical Regional Integrated Energy Systems in the Background of Energy Internet. *Proc. CSEE* **2016**, *36*, 3292–3306.
4. Chen, B.; Liao, Q.; Liu, D.; Wang, W.; Wang, Z.; Chen, S. A comprehensive evaluation index and method for regional integrated energy system. *Autom. Electr. Power Syst.* **2018**, *42*, 174–182.
5. Moeini-Aghtaie, M.; Abbaspour, A.; Fotuhi-Firuzabad, M.; Hajipour, E. A decomposed solution to multiple-energy carriers optimal power flow. *IEEE Trans. Power Syst.* **2014**, *29*, 707–716. [CrossRef]
6. Rastegar, M.; Fotuhi-Firuzabad, M.; Lehtonen, M. Home load management in a residential energy hub. *Electr. Power Syst. Res.* **2015**, *119*, 322–328. [CrossRef]
7. Herrando, M.; Markides, C.N.; Hellgardt, K.A. UK-based assessment of hybrid PV and solar-thermal systems for domestic heat-supply and power: System performance. *Appl. Energy* **2014**, *122*, 288–309. [CrossRef]
8. Li, Y.; Wu, M.; Zhou, H.; Wang, W.; Wang, D.; Ge, L. Discussion on several problems of regional multi-energy system based on all-energy flow model. *Power Syst. Technol.* **2015**, *39*, 2230–2237.
9. Kuosa, M.; Kontu, K.; Mäkilä, T.; Lampinen, M.; Lahdelma, R. Static study of traditional and ring networks and the use of mass flow control in district themaling applications. *Appl. Ther. Eng.* **2013**, *54*, 450–459. [CrossRef]
10. Li, Y.; Xun, J.; Cao, H.; Gao, C.; Zhang, X.; Zhang, J.C.W. Distribution Network Planning Strategy Based on Integrated Energy Collaborative Optimization. *Power Syst. Technol.* **2018**, *42*, 1393–1400.
11. Bai, M.; Wang, Y.; Tang, W.; Wu, C.; Zhang, B. Day-time optimization scheduling of interval integrated energy system based on interval linear programming. *Power Syst. Technol.* **2017**, *41*, 3963–3970.
12. Hu, R.; Ma, J.; Li, Z.; Lu, Q.; Zhang, D.; Qian, X. Optimal Configuration and Applicability Analysis of Distributed Cogeneration System. *Power Syst. Technol.* **2017**, *41*, 418–425.
13. Liu, D.; Ma, H.; Wang, B.; Gao, W.; Wang, J.; Yan, B. Operational optimization of regional integrated energy system with combined themal and power supply and energy storage. *Autom. Electr. Power Syst.* **2018**, *42*, 113–120.
14. Gu, W.; Lu, S.; Wang, J.; Yin, X.; Zhang, C.; Wang, Z. Multi-regional integrated energy system heat-supply network modeling and system operation optimization. *Proc. CSEE* **2017**, *37*, 1035–1046.
15. Wang, J.; Gu, W.; Lu, S.; Zhang, C.; Wang, Z.; Tang, Y. Collaborative Planning of Multi-region Integrated Energy System Based on Heat-supply network Model. *Autom. Electr. Power Syst.* **2016**, *40*, 17–24.
16. Xiao, X.; Kan, W.; Yang, Y.; Zhang, S.; Xiao, Y. Superstructure optimization configuration of combined energy storage system. *Proc. CSEE* **2012**, *32*, 8–15.
17. Wu, H.; Wang, D.; Liu, X. Strategy Evaluation and Optimization Configuration of Solar Cooling, Cogeneration System. *Power Syst. Autom.* **2015**, *39*, 46–51.
18. Gu, W.; Wu, Z.; Bo, R.; Liu, W.; Zhou, G.; Chen, W.; Wu, Z. Modeling, planning and optimal energy management of combined cooling, themaling, and power microgrid: A review. *Int. J. Electr. Power Energy Syst.* **2014**, *54*, 26–37. [CrossRef]
19. Fumo, N.; Chamra, L.M. Analysis of combined cooling, themaling, and power systems based on source primary energy consumption. *Appl. Energy* **2010**, *33*, 96–100.
20. Jing, Y.; Bai, H.; Zhang, J. Multi-objective optimization design and operation strategy analysis of solar cooling, themaling and power supply system. *Proc. CSEE* **2012**, *32*, 82–87.
21. Lo, C.E.; Borelli, D.; Devia, F.; Schenone, C. Future distributed generation: An operational multi-objective optimization model for integrated small scale urban electrical, thermal and gas grids. *Energy Convers. Manag.* **2017**, *143*, 348–359. [CrossRef]
22. Zhou, R.; Li, S.; Chen, R.; Li, H.; Yang, Y.; Chen, Y. Multi-objective Scheduling of Carbon Emissions Trading in Themal and Cold Electricity Using Fuzzy Self-Correction Particle Swarm Optimization Algorithm. *Proc. CSEE* **2014**, *34*, 6119–6126.
23. Qiu, J.; Dong, Z.; Zhao, J.; Meng, K.; Zheng, Y.; Hill, D.J. Low carbon oriented expansion planning of integrated gas and power systems. *IEEE Trans. Power Syst.* **2015**, *30*, 1035–1046. [CrossRef]
24. Wang, Z.; Liu, Y.; Tang, Y.; Gu, W.; Wu, X. Capacity allocation of CO_2 emissions considering CO_2 emissions. *Proc. CSU-EPSA* **2017**, *29*, 104–110.

25. Lorestani, A.; Ardehali, M.M. Optimal integration of renewable energy sources for autonomous tri-generation combined cooling, heating and power system based on evolutionary particle swarm optimization algorithm. *Energy* **2018**, *145*, 839–855. [CrossRef]
26. Yousefi, H.; Ghodusinejad, M.H.; Noorollahi, Y. GA/AHP-based optimal design of a hybrid CCHP system considering economy, energy and emission. *Energy Build.* **2017**, *138*, 309–317. [CrossRef]
27. Dong, X.; Quan, C.; Jiang, T. Optimal Planning of Integrated Energy Systems Based on Coupled CCHP. *Energies* **2018**, *11*, 2621. [CrossRef]
28. Ma, H.; Wang, B.; Gao, W.; Liu, D.; Sun, Y.; Liu, Z. Optimal Scheduling of an Regional Integrated Energy System with Energy Storage Systems for Service Regulation. *Energies* **2018**, *11*, 195. [CrossRef]
29. Wang, Y.; Yu, H.; Yong, M.; Huang, Y.; Zhang, F.; Wang, X. Optimal Scheduling of Integrated Energy Systems with Combined Heat and Power Generation, Photovoltaic and Energy Storage Considering Battery Lifetime Loss. *Energies* **2018**, *11*, 1676. [CrossRef]
30. Yang, L.; Zhao, X.; Li, X.; Yan, W. Probabilistic Steady-State Operation and Interaction Analysis of Integrated Electricity, Gas and Heating Systems. *Energies* **2018**, *11*, 917. [CrossRef]

© 2018 by the authors. Licensee MDPI, Basel, Switzerland. This article is an open access article distributed under the terms and conditions of the Creative Commons Attribution (CC BY) license (http://creativecommons.org/licenses/by/4.0/).

Article

Integrated Energy Transaction Mechanisms Based on Blockchain Technology

Shengnan Zhao, Beibei Wang *, Yachao Li and Yang Li

School of Electrical Engineering, Southeast University, Nanjing 210096, China; 230159182@seu.edu.cn (S.Z.); 220172697@seu.edu.cn (Y.L.); li_yang@seu.edu.cn (Y.L.)
* Correspondence: wangbeibei@seu.edu.cn; Tel.: +86-138-1380-0784

Received: 31 July 2018; Accepted: 7 September 2018; Published: 12 September 2018

Abstract: With the rapid development of distributed renewable energy (DRE), demand response (DR) programs, and the proposal of the energy internet, the current centralized trading of the electricity market model is unable to meet the trading needs of distributed energy. As a decentralized and distributed accounting mode, blockchain technology fits the requirements of distributed energy to participate in the energy market. Corresponding to the transaction principle, a blockchain-based integrated energy transaction mechanism is proposed, which divides the trading process into two stages: the call auction stage and the continues auction stage. The transactions among the electricity and heat market participants were used as examples to explain the details of the trading process. Finally, the smart contracts of the transactions were designed and deployed on the Ethereum private blockchain site to demonstrate the validity of the proposed transaction scheme.

Keywords: blockchain; decentralized market; integrated energy transaction; transaction scheme design

1. Introduction

With the promotion of technologies for distributed renewable energy (DRE) generators and demand response (DR) programs, the boundaries between generators and consumers are becoming blurred. Jeremy Rifkin presented the concept of energy interconnection, the cores of which are renewable distributed energy and energy internet that would enable access to distributed energy and fair trade [1]. This concept would also enable more forms of demand-side energy resources, such as heat and gas, to participate in the market. Traditional electricity transactions will develop into integrated energy resource (IER) transactions, which makes the transactions more complicated and the management more difficult [2]. The direction of energy delivery is vital to enhance efficiency and decrease energy consumption, such how to handle the first mile and last mile delivery logistics [3]. Therefore, designing an efficient trading scheme is vital to facilitate the development of the energy internet. The management of transactions can be divided into two primary categories: centralized and decentralized [4]. The centralized transactions may have such issues as high operating costs, excessive time consumption, and security problems. To arrange scheduling instructions, the trading center has to collect all information regarding the distributed energy, which will lead to personal privacy concerns [5]. In addition, when the number of transaction users increase, the data information will increase geometrically, which will increase the difficulty of scheduling resources in real time. Therefore, several scholars have presented the idea of decentralizing the distributed energy transactions [6].

The preconditions for the distributed energy trading were analyzed [7], which included the real-time information exchange, the self-optimizing strategy selection, the automatic transaction settlement, and the trading platform decentralization for the local energy internet. In [8], a fully decentralized microgrid platform, "Overgrid", is presented in order to carry out peer-to-peer (P2P) transactions. An automated DR program is presented in [9], which is not fully decentralized, since the transactions are considered at the level of energy aggregators, together with being not for each individual part. In both [10,11], the multiagent system is adopted to realize grid decentralization. Each of the energy consumption devices is controlled by the agent that may respond to the signals from the network. Also, in [12], the action policy put forward by each agent (termed as "scheduler" in this paper) becomes more complex, meanwhile considering more traffic issues. The distributed optimization algorithms are also employed for the purpose of decentralizing the power grid dispatching, for instance [13]. A decentralized price-based DR system is presented in [14], wherein, the price signal is adjusted by the difference in the supply and demand, and the users can adjust their demands in accordance with the signal.

Thereafter, blockchain has been considered as one of the emerging technologies, which can be employed for developing a platform for the decentralized energy transactions [15]. Considering the blockchain technology's characteristics of being open, decentralized, transparent, and tamper-resistant, it has the potential to effectively improve the efficiency of transactions, in addition to ensuring transaction security [7]. Study [16] concentrates on security issues for energy interactions and information in electric vehicles cloud and edge computing, and [17] proposes a framework based on hierarchy and distributed trust to maintain security and privacy. Solutions, methodologies and technical considerations for addressing issues of data protection, security and privacy are described in [18]. Meanwhile, blockchain is conducive to effectively improve the efficiency of transactions. In comparison with other methods, the main benefits associated with the blockchain adoption for the decentralized energy trading are that a third-party intermediary (for instance, a distribution system operator) is not required for the management and security of the energy transactions; and different kinds of energies (electricity, heat, etc.) can be freely traded [5], catering to the requirements of IER trading. Currently, blockchain technology has been thoroughly studied for its basic principles, characteristics, and applications in finance, energy and other fields [19–21]. Some research works have been carried out addressing the technical details of blockchain in order to scale up both the volume and speed of transaction by adopting the research results of cloud storage [22,23]. The research dealing with the applications of blockchain technology in energy transactions can be divided into three categories. The first category is the feasibility analysis, concentrating on the application possibilities and its capability with the trading mechanisms, for instance, renewable energy transactions, ancillary services, and large consumer's direct power trading [24–26]. The second category involves the research on the transaction process design and the smart contract design, for instance, automatic demand response, EV charging, and DRE's P2P transactions in the distribution network [27–31]. The third category deals with analysing the effect on the retail electricity market if the blockchain is widely utilized in the distributed power transactions [32]. However, the current trading mode lacks consideration for the multiple energy types as well as multilateral trading parties in local IER transactions.

This study applied blockchain to the transactions among local IER resources. The idea of blockchain-based decentralized trading was put forward for the realization of the coordinated allocation of local IER resources. Corresponding to the transaction principles, a blockchain-based integrated energy transaction mechanism is proposed, dividing the trading process into two stages: the call auction stage and the continues auction stage. We designed the smart contracts of the users' energy demands, the trading matchmaking, and the trading settlement, which were deployed on a private chain on the Ethernet for the simulation of the transaction process. The results can promote the development of a blockchain-based IER trading protocol as well as the engineering application of IER transactions.

The organization of this paper is as follows. Section 2 provides introduction to the blockchain technology and its applicability to the distributed energy transactions. Section 3 puts forward the decentralized IER transaction mechanism. Section 4 presents both the functions and elements of the smart contracts built and deployed on the Ethernet. Section 5 presents a simple case study, aimed at validating the transaction mechanism proposed.

2. Blockchain and Decentralized Energy Transactions

2.1. Blockchain and Smart Contracts

Blockchain is a chain structure in which all transaction data are packed into blocks, and the blocks are connected in chronological order [30]. Using the technology of the asymmetric encryption, the Merkel tree, and the proof of work consensus mechanism, the transaction data can be transparent, non-destructive, and traceable. The essence of a blockchain is a decentralized distributed database. Compared with the traditional database, blockchain has the advantage that the data are tamper-resistant; therefore, information is safe. However, if blockchain is only applied to data storage, its function is limited. Therefore, it is proposed to combine blockchain with smart contracts to achieve more complex functions.

A smart contract is a set of digital programs that prescribes the rights and obligations to the consumer that is automatically executed by the computer system [15]. When an agent executes a contract locally, the other agents will update the contract's content to reach a new consensus. Thus, a credible partnership can be established in a distributed system without central supervision. The blockchain is the supporting technology of smart contracts and the Ethernet is the most widely used open source blockchain platform [33]. However, for the mass adoption of this technology, volume and speed of transaction processing need to be scaled up. There is presently an unresolved tension between scalability, security, and decentralization concerns, as only two of them at a time can (so far) be addressed satisfactorily. To address the trilemma, several solutions have been proposed, such as off-chain interaction technology.

2.2. Off-Chain Interaction Technologies

Each action on the main-chain is packed and recorded into a block, which is time-consuming and limits the transaction speed. To improve the speed of the transaction processing on blockchain, off-chain technologies that build onto the main-chain are proposed. Base-level protocols do not need to change. Instead, they exist as smart contracts on the Ethereum site that interact with off-chain software. The function of a public blockchain can be expanded such that the security and irreversibility are ensured.

The typical "off-chain" technologies include state channels, Plasma, and Truebit [34]. State channels and Plasma allow interactions outside the blockchain such that the platform will have an increased transaction throughput. Truebit completes the complex calculations in the smart contracts that are off-chain. This reduces the cost of executing smart contracts, i.e., gas in Ethereum, while completing those operations with costly computation expenses or those that cannot be completed on-chain.

Off-chain technologies allow transactions to achieve a balance among speed, irreversibility, and costs by sacrificing a little decentralization. Therefore, blockchain may be applied to more extensive and complex trading modes.

2.3. Decentralized Transactions of Integrated Energy

Distributed exchange models were first applied to the financial field. With an increase in the types of tokens, the circulation among them became more difficult. Centralized exchanges are confronted with multiple risks, such as network security and protection. To solve these problems, the open source community has completed a series of attempts to build distributed exchanges, such as EtherDelta, 0x protocol, Loopring and Kyber Network. The core of the distributed exchanges is a decentralized trading protocol, which is based on blockchain [35]. This study uses Loopring as an example to explain the exchanges' working mechanisms. In Loopring, a user authorizes the matchmaking contract to transfer the tokens to a smart contract. The exchange searches for a set of orders that can be matched and transfers the matching results to the smart contract's address after the user signs in, then the order can be matched off-chain and the asset is secured [36]. By using the off-chain technologies, the distributed exchanges can complete complex deals.

The distributed exchanges on blockchain can meet the demand which integrated energy requires in market transactions. First, the characteristics of blockchain, such as equality and decentralization, can support the numerous peer-to-peer transactions among the distributed energy resources. Second, the managers of the energy subsystems are different, but blockchain provides multiple users with a decentralized and trustless power market to reduce transaction costs. Third, smart contracts provide a platform for the implementation of advanced transactions [37]. As the decision-making mode of the energy system gradually transforms to the distributed mode, the users' decisions and transactions can be written into contracts to ensure efficient decision-making and fair transaction.

3. Transaction Framework of Distributed Integrated Energy Trading

In the distributed energy trading framework designed in this paper, we assumed that each user had a blockchain account with a pair of keys, i.e., a public key and a private key [4], and that smart meters were available to upload power data to the network. An energy usage plan should be determined before the trading session. The participants in the distributed energy trading must authorize the exchange to transfer a set number of tokens from their blockchain accounts to multi-signature wallets. Inspired by the mechanism of the stock market exchange, the transaction process proposed in this paper is divided into two stages: the call auction stage and the continuous double auction stage. The schematic process is presented in Figure 1. The energy buyer authorizes the exchange so that the software could transfer a set number of tokens from the blockchain account to multi-signature wallets (signed by both the buyer and the exchange). The order contracts which contain the buyers' order information and the sellers' information can be sent to the exchange with the digital signatures by using their private key. The exchange classifies the orders provided by the buyers and the sellers, followed by combining them into order books. The form of the order books requires conforming to two principles: the same kind of energy and the same delivery time, so that the two principles could enter the settlement within the same trading period T. Thereafter, the transaction processing contract or the matchmaking trading contract will facilitate the transactions. Moreover, the settlement contract will be created and deployed to the main chain for the purpose of ensuring the assets' safety for the traders. After reaching the transaction agreement, the smart meters' data will be uploaded to the settlement smart contract automatically and the tokens will be transferred as well.

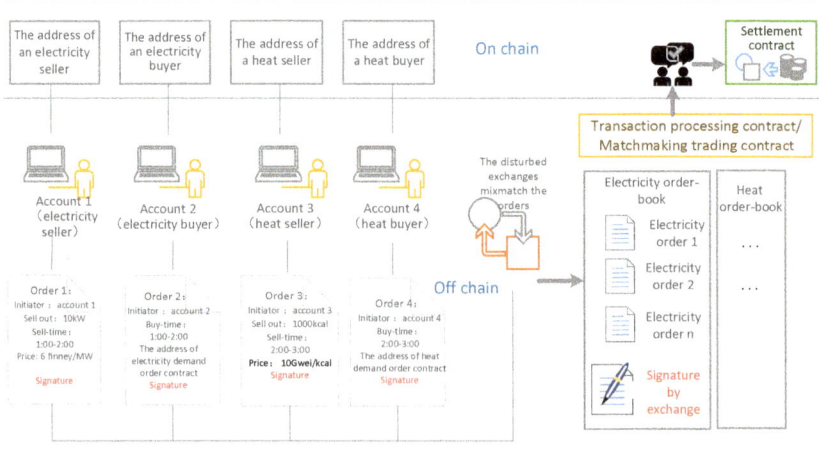

Figure 1. Process diagram of the transactions.

The parameters used in this section are given in abbreviations Table 1.

Table 1. Abbreviations.

Items	Meaning
$P_{CHP}(t)$	The electricity production of CCHP at time t
F_{CHP}	The natural gas consumed by CCHP
$\eta_{CHP,E}$	The electricity generation efficiency of CCHP
Q_{CHP}	The recovered heat energy of CCHP
$\eta_{CHP,H}$	The heat recovery efficiency of CCHP
P_{Gi}	The energy amount provided by energy seller i
$p_{Gi}^{\min}$	The price floor of energy seller i
P_{Dj}	The energy demand of the buyer j
σ	The difference between energy demand and supply
λ^0	The initial quoted price broadcasted to buyers
λ^k	The energy price in kth iteration
ρ	The step length set in the iterative process
α_i, ω_i	The coefficients of the electricity user i's utility function
$U_{Di}(P_{Di})$	The benefit of the electricity user i if P_{Di} is purchased
π_{Di}	The profits of electricity user i
$E_{th,i}$	The ith energy seller's profit
P'_{Gi}	The actual energy amount provided by the ith energy seller
p_T	The energy's price
τ	The penalty parameter if the actual energy is too much or too little
$\tau_b(t)$	The buyer's appraisal price in the tth iteration of trading
$\tau_s(t)$	The seller's appraisal price in the tth iteration of trading
p_{call}	The energy price got form call auction
$p_{Obid}(t)$	The optimal purchase price in the tth iteration of trading
$p_{Oask}(t)$	The optimal selling price in the tth iteration of trading
$\eta_b, \eta_s \in (0,1]$	the increment rates of the buyer's and seller's quotation

3.1. Call Auction Stage

During the call auction stage, the seller and buyer will make disposable matchmaking transactions providing guiding prices for various types of energy at each period in a day. The matchmaking transaction may be used as a reference for the quoted price during the double auction stage. It should be noted that while there are many ways to hold a call auction, this paper uses only one of them.

Besides the proceeding transaction process, the process in call stage has its own features:

(1) In call stage, a demand order contract address is required for each buyer's order; and the seller i writes its supply p_{Gi}^T and price floor $p_{Gi}^{T,\min}$ during the trading period T into its bid order. For presentation purpose, the superscript letter T of parameters are omitted in the following.

(2) All the information is sent to the exchange, which is responsible for classifying the orders into groups based on the energy types and trading time. The orders in the same order-book will be sent to one transaction processing smart contract, and the contract calculates the maximum value of the price floors submitted by all sellers, followed by delivering the price signal to the demand order contract. Demand (obtained from the demand order contract) and supply are followed by adjusting the price signal in accordance with the Formulas (1)–(3).

(3) In each interaction, the demand order contract modifies the amount of demand in accordance with its utility function. In the meantime, the supply is calculated at the quoted price signal. The interactions end until Formula (4) is satisfied.

$$\sigma = \sum_{i=1}^{N_G} P_{Gi} - \sum_{j=1}^{N_L} P_{Dj} \tag{1}$$

$$\lambda^0 = \max(p_{Gi}^{\min}) \tag{2}$$

$$\lambda^{k+1} = \lambda^k + \rho \cdot (\sigma / \sum_{i=1}^{N_G} P_{Gi}) \qquad (3)$$

$$\sigma \leq \varepsilon \qquad (4)$$

We use the power buyers as an example. The electricity users' utility function [38] and profit function are shown as Formulas (5) and (6). Meanwhile, there are constraints on the users' purchase quantity, per Formula (7).

$$U_{Di}(P_{Di}) = \begin{cases} -\alpha_i P_{Di}^2 + \omega_i P_{Di} & P_{Di} \leq \omega_i/2\alpha_i \\ \omega_i^2/4\alpha_i & P_{Di} > \omega_i/2\alpha_i \end{cases} \qquad (5)$$

$$\pi_{Di} = U_{Di}(P_{Di}) - \lambda P_{Di} \qquad (6)$$

$$P_{Di}^{min} \leq P_{Di} \leq P_{Di}^{max} \qquad (7)$$

Based on the principle of profit maximization, the demand can be calculated according to Formula (8):

$$P_{Di} = \begin{cases} (\omega_i - \lambda)/2\alpha_i & P_{Di}^{min} \leq \omega_i/2\alpha_i \\ P_{Di}^{min} & P_{Di}^{min} > \omega_i/2\alpha_i \end{cases} \qquad (8)$$

Therefore, according to price λ_i^{k+1}, corresponding P_{Di} is provided by electricity users. This process can be written as a contract, which is deployed on the side-chain managed by the exchange, and executed off the main chain.

(4) When Formula (4) is satisfied, pre-transaction orders are generated. The exchange adjusts the orders as per the amounts of another type of energy provided by the multi-energy sellers.

For instance, the sellers, capable of supplying two or more types of energy, require special handling. The pre-transaction orders should be adjusted in accordance with their operation mode. For instance, combined cooling, heating and power (CCHP) follows the supply formula presented hereunder:

$$P_{CHP}(t) = F_{CHP}(t)\eta_{CHP,E} \qquad (9)$$

$$Q_{CHP}(t) = F_{CHP}(t)(1 - \eta_{CHP,E})\eta_{CHP,H} \qquad (10)$$

The operation mode should be determined through the determination of either the power by heat or the heat by power. In case of determination of power by heat, the pending transaction will be based on the power. The heat energy supply is subsequently calculated in accordance with the power transaction volume. The exchange will replace the heat seller having the same amount, but at a higher price in the same period, and vice versa.

(5) The transaction processing smart contract creates objects for the settlement contracts based on the consensus of the transaction. Following the settlement, the transaction amount is calculated in accordance with the Formula (11), which is based on the seller's smart meter's data. The money $E_{th,i}$ in the multi-signature wallet is subsequently transferred to the seller by the means of state channels:

$$E_{th,i} = P'_{Gi} \cdot p_T \cdot \tau \qquad (11)$$

3.2. Continuous Double Auction Stage

After the call auction, traders who do not have a deal and new traders can hold continuous double auctions. During the trading cycle, buyers and sellers can submit their quotations at any time. The buyers' prices are ranked from high to low and the highest quotation is the optimal purchase price. The sellers' prices are ranked from low to high and the lowest quotation is the optimal selling price. The buyer who has the highest quotation can match the seller who has the lowest quotation. When the optimal purchase price is higher than or equal to the optimal selling price (the price margin is less than or equal to zero), a deal can be made, as Figure 2 shows. Actual price is the average price of the two quotations. If the prices are identical, the transactions are completed according to the quotations' submission time. During the matchmaking process, the matchmaking smart contracts need to update the status of each quotation (withdrawal or addition) in real time.

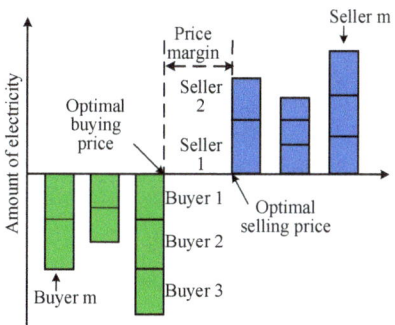

Figure 2. Matching process of continuous double auction.

During the auction, market participants can adjust their quotations based on market information for the next cycle of trading until the total amount of electricity is sold out or the trading time is cut off. After completing the matching transactions, the two sides create the settlement smart contract and transfer the tokens.

Unlike the centralized approach, in the continuous double auction stage, the individuals participating in energy trading are interested in maximizing its utility (energy to be requested) without concern for the interests of others. Every participant can submit bidding to the matching contract at each trading interval. In this paper, we assume all participants adopted the AA strategy [39], which adjusts the price according to the marketing environment.

Before entering the market, each buyer and seller calculate an appraisal price according to the expected purchasing price, production cost, deal price from the centralized competitive bidding stage and trading willingness [39]. Thus, the buyer's quoted price $p_b(t)$ and the seller's quoted price $p_s(t)$ are calculated according to Formulas (12) and (13), respectively:

$$p_b(t) = \begin{cases} p_{call}, & t = 1 \\ p_{Obid}(t) + \eta_b(\tau_b(t) - p_{Obid}(t)), & t \geq 2 \end{cases} \quad (12)$$

$$p_s(t) = \begin{cases} p_{call}, & t = 1 \\ p_{Oask}(t) - \eta_s(p_{Oask}(t) - \tau_s(t)), & t \geq 2 \end{cases} \quad (13)$$

In the first iteration of trading, the results of the call auction are used as the optimal purchase price and the optimal selling price. In the following transactions, if the sellers and the buyers have not completed a deal, the buyer and the seller can adjust the quotation according to the base price and the current optimal selling price.

4. Formulation of Smart Contracts

Off-chain interaction technology is still in the initial stage of development; even the most mature application, Lightning Network, which promises to ease network congestion by moving transactions off the main blockchain, while it has not been suitable for high value payment system in light of security, only realizes the micropayment function in the network. Therefore, we will only discuss the contract related to the transaction mechanism in this paper. This section will discuss the construction of smart contracts by using the transaction between the distributed generators and electricity users as an example.

The transaction process involves five smart contracts in chronological order, including the transaction processing in the call auction stage, the determination of each user's demand, the adjustment of scheduled orders, the matchmaking in the continuous double auction stage and the transaction settlement.

4.1. The Transaction Processing Contract

The input of this contract is the supply amount during a certain period and the corresponding price floor submitted by the sellers. The smart contract will record all orders and calculate the total power supply at the power price obtained by the iterative double auction. The smart contract will subsequently adjust the price and send the new price to the address of the customer's demand order contract. The outputs include new electricity price (being sent back to the user's wallet when the iterative termination criteria are not satisfied) and the trading volume of each seller (output when the iterative termination criteria are satisfied).

When electricity supply equals the demand, the contract will check whether the heat orders have been processed. If so, the contract will do two jobs. One is to calculate the heat supply according to the quantity of electricity provided by the CCHP and then transmit the data to the adjustment contract of the scheduled heat sellers. The other is to adjust the scheduled sellers according to the power data from the adjustment contract of the scheduled heat sellers and replace the same amount of electricity bought from the generators with the highest price floor. The basic elements of the contract are provided in Table 2.

Table 2. Basic elements of a transaction processing contract.

The Parameters' Name	Type	Meaning
Seller address	Address	Account address of sellers
Electricity supply	Uint	Determined by sellers
Heat supply	Uint	Determined by DG type and the DG's electricity supply
Price floor	Uint	Determined by sellers
The delivery time	Int256	Determined by sellers
Demand order contract's address	Address	The demand order contract's address of the Order Set
Price	Uint	Initial value is determined by the average value of price floor. Final value is determined by iterative double auction.

4.2. Demand Order Contract of Electricity Buyers

The buyers should transfer a deposit to the multi-signature wallet's address according to the local electricity retail price. This part of the work is done before the buyer joins the market (building a status channel or joining the side-chain managed by the exchange). The input of this contract is the energy type, energy amount, and delivery time of sellers and buyers. The output includes the buyer's address, the seller's address, the energy amount and the energy price.

The function of this contract is to calculate user demand according to the price provided by the transaction processing contract. The basic elements of the contract are shown in Table 3.

Table 3. Basic elements of a demand order contract.

The Items of Contract	Type	Meaning
Buyer address	Address	Account address of the buyer
Total amount	Int256	Determined by the buyer's strategy
The delivery time	Int256	Determined by the buyer

4.3. The Matchmaking Trading Contract

The matchmaking trading contract checks the delivery time and the energy type of the sellers and the buyers and classifies buyers and sellers of the same delivery time according to bid price. The contract then matches the sellers and buyers by the transaction mechanism presented in Section 3.2.

The basic elements of the contract are shown in Table 4.

Table 4. Basic elements of a matchmaking trading contract.

The Parameters' Name	Type	Meaning
Seller address	Address	Account address of sellers
Seller's energy type	Uint	Determined by sellers, use 0 for electricity, 1 for heat.
Seller's energy quantity	Uint	Determined by sellers
Seller's energy price	Uint	Determined by sellers
Sellers's delivery time	Uint	Determined by sellers
Buyer's energy type	Uint	Determined by buyers, use 0 for electricity, 1 for heat.
Buyer's energy quantity	Uint	Determined by buyers
Buyer's energy price	Uint	Determined by buyers
Buyer's delivery time	Uint	Determined by buyers

4.4. The Settlement Contract

By means of the settlement contract, the funds can be automatically transferred according to the pre-agreed terms and deal terms after the delivery time of the transaction. The basic elements of the contract are shown in Table 5.

Table 5. Basic elements of a settlement contract.

The Parameters' Name	Type	Meaning
Seller address	Address	Account address of sellers
Buyer address	Address	Account address of buyers
Target energy amount	Uint	Determined by the transaction result
Price	Uint	Determined by double auction
Actual energy amount	Uint	Determined by the seller's electricity supply
Penalty parameter	Uint	Determined by pre-agreed terms
The delivery time	Int256	Determined by sellers

The energy amount supplied by the sellers is submitted by smart meters. The settlement rules should be clarified in the market access agreements signed before the smart contract is created. After the delivery period, the electric data will be transferred to the settlement contract. The contract will be executed and the tokens will be transferred from the multi-signature wallet to the multi-signature wallet of the seller and the exchange account.

5. Case Study

To verify the effectiveness of the transaction mechanism proposed in this paper, the smart contracts for the distributed transaction were released on the test network of the Ethernet for simulation. There were five electricity buyers, in addition to four generators, merely providing electricity and two CCHP in this test. One of the two CCHP works to determine power by heat operation mode (this CCHP joins in the heat transaction processing contact) and other works in the determination of the heat by power operation mode (this CCHP joins the electricity transaction processing contract). We employ the Remix platform to test the smart contracts.

5.1. Deploying the Demand Order Contract of Electricity Buyers

The demand order contract consists of a constructor, a power calculation function, a data transfer function, and auxiliary functions. Constructor function has the same name as the contract. With the aid of this function, the contract can be deployed preliminarily and the initial variable can be initialized. The inputs of constructor are the parameters of the electricity buyers' demand strategies, which are shown in Table 6. These buyers create the demand orders and send them to the exchange address. This paper places all user data into one contract for demonstration purposes.

Table 6. Parameters of users' demand function.

No.	$2\alpha_i$	ω_i	P_{Gi}^{max} ($\times 100$ kW)	P_{Gi}^{min} ($\times 100$ kW)
1	0.34	42.5	2.46	0.4
2	0.33	30.6	1.85	0.5
3	0.23	33.1	2.4	0.8
4	0.32	36.7	1.8	0.7
5	0.20	40.3	2.7	1.0

We use Solidity programming language, which is a contract-oriented, high-level language designed to target the Ethereum Virtual Machine (EVM), to build the contracts. Considering that the Solidity does not support decimal numeric storage, in order to ensure the accuracy of the calculation results, the initial data are magnified. The values of the buyers' parameters are magnified 100 times as the constructor input. The operating results of the contract are shown in Figure 3, which indicate the state of execution of the contract, the address of the contract, the address of the executor, the gas value of the execution of the contract, the contract's hash value, the log, and other information.

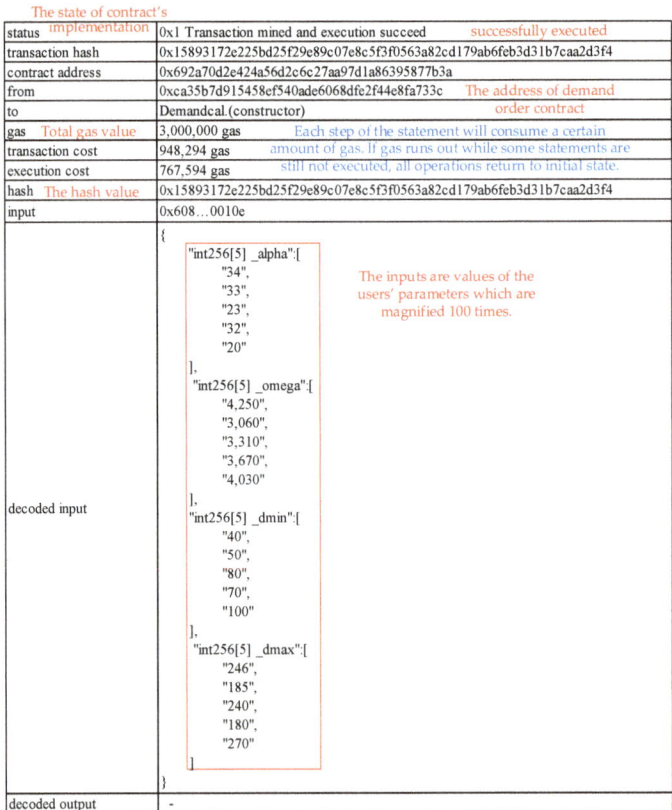

Figure 3. Input of the buyers' characteristic parameters.

The demand order contract calculates the output based on the quoted price according to Formulas (9) and (10). As shown in Figure 4, the outputs of four users from the initial quotation (11 Finney/MW, where Finney is the unit of virtual currency in Ethereum) are 920, 590, 960, 800 and 1460 kW.

Energies 2018, *11*, 2412

hash	0x5eea4eeb2b1015ceaf24f2176e61dd4a01925a2f15add2a34d2ccc592283b291
input	0x635...0044c
decoded input	{ "int256 gamma": "1,100" Initial quotation }
decoded output	{ "0": "int256[5]: 92,59,96,80,146" Four users' outputs }
logs	[]

Figure 4. Buyers' demand value at the initial quoted price.

The call function can transmit the output of each electricity buyer to the transaction processing contracts. In case of being successful, call function will return a Boolean parameter "true". The process is presented in Figure 5. An array of five integer variables, presenting the buyers' electricity demand, was transmitted from the address of the demand order contract to the transaction processing contract.

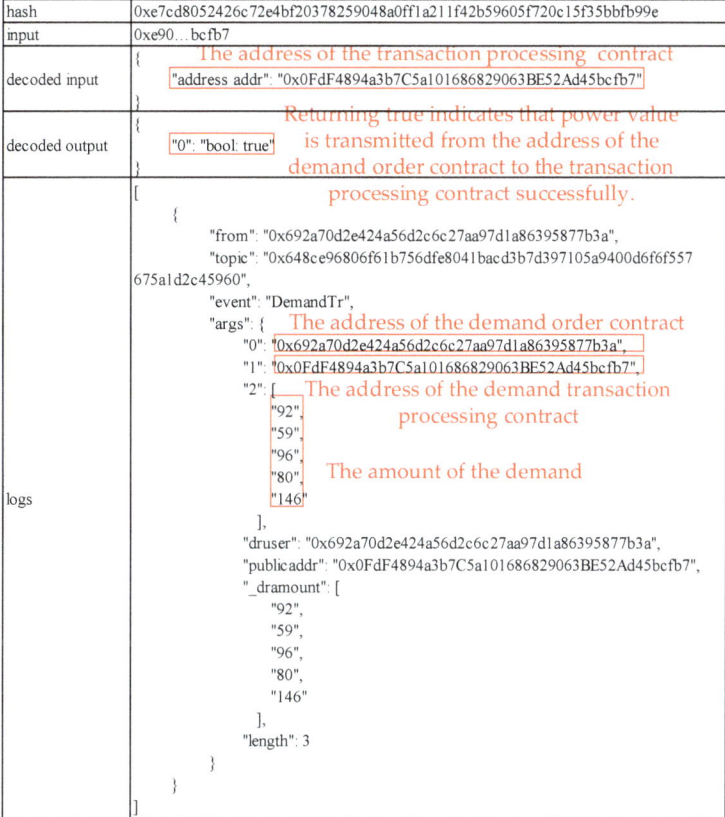

Figure 5. Transferring of buyers' demand using the initial quoted price.

5.2. Deploying the Transaction Processing Contract

The transaction processing contract can be deployed on the exchange's wallet, to which the energy buyers transfer a margin prior to trading. This contract consists of a constructor, an input interface function, call functions, a price adjustment function, a supply amount calculation function, the other kind of energy's amount calculation function, a seller readjustment function, and auxiliary functions.

The assumed seller's parameters are presented in Table 7. It requires observation that only the sellers, having the same delivery time, will be sent to the same contract.

Table 7. Sellers' parameters.

No.	Generator Type (1 for CCHP, 0 for Normal DG)	Delivery Time	P_G ($\times 100$ kW)	The Corresponding Price (Finney/MW)	P_G^{min} ($\times 100$ kW)	The Number of Bidding's Piecewise
1	1	8:00–9:00	150,100,150	11,13,15	50	3
2	0	8:00–9:00	100,50,50	9,12,14	30	3
3	0	8:00–9:00	20,15,15	5,8,11	0	3
4	0	8:00–9:00	100	6	0	1

The result of input interface is presented in Figure 6. Furthermore, the initial quotation (11 Finney/MW) is transmitted to the buyers' demand order contract, and the process is recorded in an event. By the call function, the integer data 1100 is transferred from the address of the transaction processing contract to the demand order contract, as presented in Figure 7.

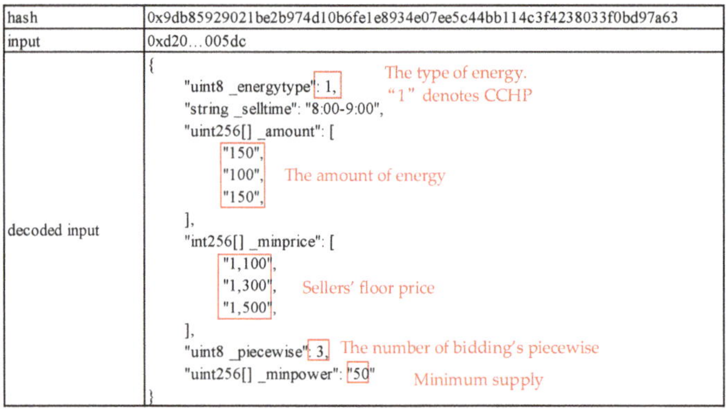

hash	0x9db85929021be2b974d10b6fe1e8934e07ee5c44bb114c3f4238033f0bd97a63
input	0xd20...005dc

decoded input	{
	 "uint8 _energytype": 1, The type of energy. "1" denotes CCHP
	 "string _selltime": "8:00-9:00",
	 "uint256[] _amount": [
	 "150",
	 "100", The amount of energy
	 "150",
	],
	 "int256[] _minprice": [
	 "1,100",
	 "1,300", Sellers' floor price
	 "1,500",
	],
	 "uint8 _piecewise": 3, The number of bidding's piecewise
	 "uint256[] _minpower": "50" Minimum supply
	 }

Figure 6. Input of sellers' parameters.

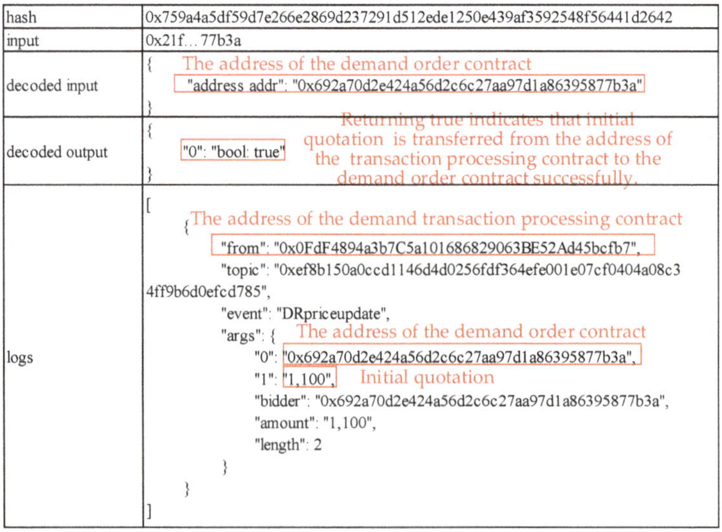

hash	0x759a4a5df59d7e266e2869d237291d512ede1250e439af3592548f56441d2642
input	0x21f...77b3a
decoded input	{ The address of the demand order contract "address addr": "0x692a70d2e424a56d2c6c27aa97d1a86395877b3a" }
decoded output	{ Returning true indicates that initial quotation is transferred from the address of "0": "bool: true" the transaction processing contract to the demand order contract successfully. }
logs	[{ The address of the demand transaction processing contract "from": "0x0FdF4894a3b7C5a101686829063BE52Ad45bcfb7", "topic": "0xef8b150a0ccd1146d4d0256fdf364efe001e07cf0404a08c34ff9b6d0efcd785", "event": "DRpriceupdate", "args": { The address of the demand order contract "0": "0x692a70d2e424a56d2c6c27aa97d1a86395877b3a", "1": "1,100", Initial quotation "bidder": "0x692a70d2e424a56d2c6c27aa97d1a86395877b3a", "amount": "1,100", "length": 2 } }]

Figure 7. Record of transferred quoted price.

The output returned by the demand order contract is [92,59,96,80,146]. Thereafter the transaction processing contract adjusts the prices in accordance with the values for Formulas (1)–(3) and the updated price is 12.21 Finney/MW. The supply calculation function calculates the electricity provided during the iterations, which is 4500 kW. Also, if the supply is not equal to demand, then proceed to the next iteration.

Following three iterations, the buyers' demands are 890, 550, 900, 760 and 1400 kW correspondingly, while the price is 12.24 Finney/MW. Subsequently, setting the balance indicator inside the contract to be "true" and checking the balance indicator of heat transaction processing contract. If the indicator is "true", then this contract will get the electricity generated by the CCHP that is scheduled by the heat transaction processing contract and send the heat generated by the CCHP which is scheduled by this contract. The way to transfer data between contracts is same to the call function mentioned above.

The sellers' scheduled energy amount is presented in Table 8. Assuming the electricity amount received from the heat transaction processing contract is 400 kW, which is provided by the CCHP "0xdd870fa1b7c4700f2bd7f44238821c26f7392148". The adjustment is recorded as an event, which is presented in Figure 8. This figure suggests that the seller "0x14723a09acff6d2a60dcdf7aa4aff308fddc160c" reduces 400 kW.

Table 8. Sellers' scheduled energy amount before adjustment.

Seller	Address	The Scheduled Amount before Adjustment
1	0xca35b7d915458ef540ade6068dfe2f44e8fa733c	150
2	0x14723a09acff6d2a60dcdf7aa4aff308fddc160c	150
3	0x4b0897b0513fdc7c541b6d9d7e929c4e5364d2db	50
4	0x583031d1113ad414f02576bd6afabfb302140225	100

5.3. Deploying the Matchmaking Trading Contract

This contract realizes the matchmaking of the sellers and the buyers in accordance with the method in the continuous double auction stage. This contract consists of a constructor, two input interface functions, a matchmaking function, a withdraw function and auxiliary functions.

The two input interface functions are used to receive the data from both the sellers and buyers, together with distributing them into order sets of different times and energy types. The interface function for the buyers is "payable" function, requiring the buyer to send its' advance deposit. The bidding of buyers and sellers is presented in Table 9. With regard to the demonstration purposes, they are assumed to bid for the same energy type and delivery time.

Table 9. The bidding of buyers and sellers.

Seller	Energy Type (0 for Electricity, 1 for Heat)	Energy Amount	Delivery Time (1 h as an Interval)	Bidding Price (×0.01 Finney)	Appraisal Price (×0.01 Finney)
seller0	0	20	8	930	600
seller1	0	80	8	1071	900
seller2	0	50	8	1158	1100
Seller3	0	50	8	1211	1200
buyer0	0	40	8	1112	1112
buyer1	0	30	8	1215	1215
buyer2	0	70	8	1075	1075
buyer3	0	50	8	1180	1180

```
logs   [
            {
                "from" : "0x692a70d2e424a56d2c6c27aa97d1a86395877b3a",
                "topic" : "0xf616546b8fe1b9ad8cac8aaaeac9c1813df435e5384ef9e2026877342f16da0d",
                "event": "Afteradjust",
                "args": {
                    "0": "0xCA35b7d915458EF540aDe6068dFe2F44E8fa733c",
                    "1": "150",              The address of seller 1
                    "_seller": "0xCA35b7d915458EF540aDe6068dFe2F44E8fa733c",
                    "_energyamount": "150",   The scheduled energy amount after
                    "length": 2                                  adjustment
                }
            }
            {
                "from" : "0x692a70d2e424a56d2c6c27aa97d1a86395877b3a",
                "topic" : "0xf616546b8fe1b9ad8cac8aaaeac9c1813df435e5384ef9e2026877342f16da0d",
                "event": "Afteradjust",
                "args": {                         The address of seller 2
                    "0": "0x14723A09ACff6D2A60DcdF7aA4AFf308FDDC160C",
                    "1": "110",
                    "_seller": " 0x14723A09ACff6D2A60DcdF7aA4AFf308FDDC160C",
                    "_energyamount": "110",  The scheduled energy amount after
                    "length": 2                                   adjustment
                }
            }
            {
                "from" : "0x692a70d2e424a56d2c6c27aa97d1a86395877b3a",
                "topic" : "0xf616546b8fe1b9ad8cac8aaaeac9c1813df435e5384ef9e2026877342f16da0d",
                "event": "Afteradjust",
                "args": {
                    "0": "0x4B0897b0513fdC7C541B6d9D7E929C4e5364D2dB",
                    "1": "50",
                    "_seller": " 0x4B0897b0513fdC7C541B6d9D7E929C4e5364D2dB",
                    "_energyamount": "50",
                    "length": 2
                }
            }
            {
                "from" : "0x692a70d2e424a56d2c6c27aa97d1a86395877b3a",
                "topic" : "0xf616546b8fe1b9ad8cac8aaaeac9c1813df435e5384ef9e2026877342f16da0d",
                "event": "Afteradjust",
                "args": {
                    "0": "0x583031D1113aD414F02576BD6afaBfb302140225",
                    "1": "100",
                    "_seller": " 0x583031D1113aD414F02576BD6afaBfb302140225",
                    "_energyamount": "100",
                    "length": 2
                }
            }
            {
                "from" : "0x692a70d2e424a56d2c6c27aa97d1a86395877b3a",
                "topic" : "0xf616546b8fe1b9ad8cac8aaaeac9c1813df435e5384ef9e2026877342f16da0d",
                "event": "Afteradjust",
                "args": {
                    "0": "0xdD870fA1b7C4700F2BD7f44238821C26f7392148",
                    "1": "40",               The address of CCHP
                    "_seller": " 0xdD870fA1b7C4700F2BD7f44238821C26f7392148",
                    "_energyamount": "40",   The scheduled energy amount after
                    "length": 2                                   adjustment
                }
            }
        ]
```

Before adjustment, seller 2 has 1500kW. After adjustment, seller 2 has 1100kW. Thus, seller 2 reduces 400kW.

Before adjustment, CCHP has 0kW. After adjustment, CCHP has 400kW. Thus, CCHP gains 400kW.

Figure 8. Record of the sellers' scheduled amount after adjustment.

The matchmaking function finds the highest buyers' bidding price and the lowest sellers' bidding price, followed by comparing them and calculating the transaction price and amount. If the bidding is executed in all, the contract will set the state indicator of the buyer or the seller as "true" in order to exclude them from the order set. The transaction in the first round is recorded in the event, called "Transaction log", as presented in Figure 9. The Seller0 sells Buyer1 200 kW at 10.72 Finney/MW, and the state indicator of Seller0 is set to be "true", implying that the Seller0's energy has been sold out.

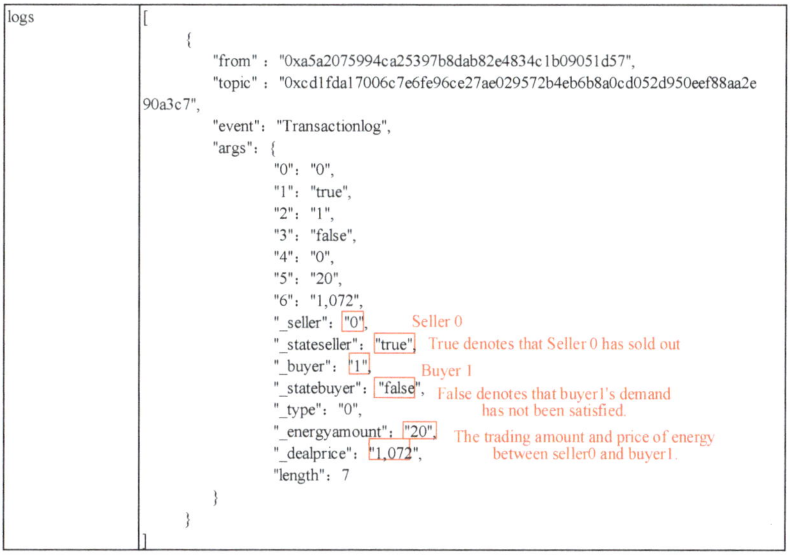

```
logs   [
         {
            "from" : "0xa5a2075994ca25397b8dab82e4834c1b09051d57",
            "topic" : "0xcd1fda17006c7e6fe96ce27ae029572b4eb6b8a0cd052d950eef88aa2e90a3c7",
            "event": "Transactionlog",
            "args": {
               "0": "0",
               "1": "true",
               "2": "1",
               "3": "false",
               "4": "0",
               "5": "20",
               "6": "1,072",
               "_seller": "0",            Seller 0
               "_stateseller": "true",    True denotes that Seller 0 has sold out
               "_buyer": "1",             Buyer 1
               "_statebuyer": "false",    False denotes that buyer1's demand
               "_type": "0",                     has not been satisfied.
               "_energyamount": "20",     The trading amount and price of energy
               "_dealprice": "1,072",            between seller0 and buyer1.
               "length": 7
            }
         }
       ]
```

Figure 9. The transaction log in the first matchmaking round.

Other functions are common in the functional design of smart contracts. The heat bidding can be processed in the same manner.

Over the trading cycle, there are 5 transactions concluded in 2 rounds of trading, as presented in Table 10.

Table 10. The transactions' information over the trading cycle.

Trading Cycle	Transaction Sequence	The Transaction Parties	Energy Amount	Transaction Price (×0.01 Finney)
Round 1	1	Seller0→Buyer1	20	1072
Round 1	2	Seller1→Buyer1	10	1143
Round 1	3	Seller1→Buyer3	50	1125
Round 1	4	Seller1→Buyer0	20	1091
Round 2	5	Seller2→Buyer0	20	1112

5.4. Deploying the Transaction Settlement Contract

The transaction settlement contract primarily comprises the constructor, the funds transferring function, and auxiliary functions. This contract should be deployed on the main chain of Ethereum and the transfer of funds should be confirmed by the multiple signatures (the exchange and the buyer).

This paper takes the Seller0 and the Buyer1 in Table 8 as an example for the illustration of the process of the transaction settlement when the energy has been delivered.

The input parameters of the contractor function include "0xb868ab9cf247345f586fa0f0750 ce110c2202db3" (the address of the user), 200 kW (the target output), 10.72 Finney/MW (the transaction price).

Assuming that the actual output of the seller is 150 kW, which is insufficient, the seller faces a punishment that the contract is settled at 90% of the transaction price. The amount that the buyer requires paying to the seller is presented in Figure 10. It records the transfer of tokens between the two addresses in the form of an event.

Figure 10. The record of the amount transferred in Metamask.

These contracts were deployed on Metamask, an application allowing the users to run Ethereum dApps in a browser. The event that records the transaction's process is presented in Figure 10. This figure reveals that 1.4472 Finney has been transferred from the address of the generator to the user.

5.5. Transaction Mechanism Comparision

The social welfare and market efficiency [40] are chosen as the evaluation indicators in order to compare the mechanism put forward in this paper, the single call auction mechanism and the single continuous action mechanism. The social welfare of the proposed mechanism, the single call auction mechanism and the single continuous action mechanism is 130.88 Finney, 130.49 Finney and 121.72 Finney, correspondingly. Market efficiency is the ratio between actual social welfare and maximum social welfare 134.18 Finney (all transactions are dealt with at a time, which is unrealistic because of the renewable energy's forecast errors and uncertainty of consumers' behaviors), which are 97.25%, 97.54% and 90.71%, correspondingly.

In accordance with the principle of economics, when the market transaction price is equal to the equilibrium price, social welfare is the largest, which is the situation with maximum social welfare. In the single call auction mechanism situation, the market is cleared once in each trading cycle (for instance, once in an hour); if the seller's or buyer's delivery time is close to the bidding time, it would not have a second bidding chance. The transaction model put forward in this paper is a compromise of market efficiency and uncertainty characteristics of market participants (for instance, renewable energies and energy consumption behaviour), which achieve the best performance among the three practical mechanisms. The mechanism put forward in this paper is superior to the single call auction mechanism since the transactions made in call stage reach the beneficial maximum. In short, the transaction model has the potential to balance the market efficiency and the trading convenience.

6. Conclusions

The decentralized characteristic of blockchain technology is its high compatibility with the trading demand of integrated energy resources. This paper discussed the application of blockchain technology in the integrated energy trading and put forward a decentralized transaction scheme, including the interaction process of the smart contracts.

The transaction process put forward in this paper is divided into two stages: the call auction stage and the continuous double auction stage. Accordingly, the trading parties adjusted their bidding by monitoring the information broadcasted in the blockchain so that more freedom could be provided for them to participate in the market. The authors designed and deployed the trading smart contracts on the Ethereum website, in addition to verifying the trading process. In the study case, distributed generations and users can achieve transaction conclusion and transaction settlement without the participation of centralized institutions; besides, the characteristics of blockchain can guarantee the interests of the market participants. In comparison with the single type transaction model, the proposed mechanism finds a compromise between the market efficiency and the participants' convenience. The proposed mechanism can meet the demand of the distributed, small-scale and low-cost transaction of the integrated energy resources in micro-grid, and can be extended to the transactions for the other production forms, such as energy efficiency in supply chain and transportation solutions.

Further studies will focus on two major parts: firstly, setting up a platform to put the distributed transaction mechanism into practice, and evaluating its effectiveness and practicability. Secondly, designing the blockchain-based transaction mechanism among the virtual power plants and the distributed resources.

Author Contributions: Conceptualization, S.Z. and B.W.; Methodology, S.Z. and B.W.; Software, Y.L. (Yachao Li); Writing—Original Draft Preparation, S.Z.; Writing—Review and Editing, Y.L. (Yachao Li); Supervision, Y.L. (Yang Li).

Funding: This research was funded by National Natural Science Foundation of China with the project No.71471036 and by National Power Grid Corp Headquarter Science and Technology Project No.52110118000A.

Conflicts of Interest: The authors declare no conflict of interest.

References

1. Rifkin, J. The third industrial revolution: How lateral power is transforming energy, the economy, and the world. *Survival* **2011**, *2*, 67–68. [CrossRef]
2. Nieße, A.; Lehnhoff, S.; Tröschel, M.; Uslar, M.; Wissing, C.; Appelrath, H.J. Market-based self-organized provision of active power and ancillary services: An agent-based approach for Smart Distribution Grids. In Proceedings of the 2012 Complexity in Engineering (COMPENG), Aachen, Germany, 11–13 June 2012.
3. Bányai, T. Real-Time Decision Making in First Mile and Last Mile Logistics: How Smart Scheduling Affects Energy Efficiency of Hyperconnected Supply Chain Solutions. *Energies* **2018**, *11*, 1833. [CrossRef]
4. Aitzhan, N.Z.; Svetinovic, D. Security and privacy in decentralized energy trading through multi-signatures, blockchain and anonymous messaging streams. *IEEE Trans. Depend. Secur. Comput.* **2016**, *99*, 1–14. [CrossRef]
5. Pop, C.; Cioara, T.; Antal, M.; Anghel, I.; Salomie, I.; Bertoncini, M. Blockchain based decentralized management of demand response programs in smart energy grids. *Sensors* **2018**, *18*, 162. [CrossRef] [PubMed]
6. Yuan, Y.; Wang, F.Y. Blockchain: The state of the art and future trends. *Acta Autom. Sin.* **2016**, *42*, 481–494. [CrossRef]
7. Zeng, M.; Cheng, J.; Wang, Y.; Li, Y.; Yang, Y.; Dou, J. Primarily research for multi module cooperative autonomous mode of energy internet under blockchain framework. *Proc. CSEE* **2017**, *37*, 3672–3681. [CrossRef]
8. Croce, D.; Giuliano, F.; Tinnirello, I.; Galatioto, A.; Bonomolo, M.; Beccali, M.; Zizzo, G. Overgrid: A Fully Distributed Demand Response Architecture Based on Overlay Networks. *IEEE Trans. Autom. Sci. Eng.* **2017**, *14*, 471–481. [CrossRef]
9. Giovanelli, C.; Kilkki, O.; Seilonen, I.; Vyatkin, V. Distributed ICT Architecture and an Application for Optimized Automated Demand Response. In Proceedings of the IEEE ISGT-Europe, Ljubljana, Slovenia, 9–12 October 2017; pp. 1–6. [CrossRef]
10. Dusparic, I.; Taylor, A.; Marinescu, A.; Cahill, V.; Clarke, S. Maximizing Renewable Energy Use with Decentralized Residential Demand Response. In Proceedings of the 2015 International Smart Cities Conference, Guadalajara, Mexico, 25–28 October 2015. [CrossRef]
11. Siebert, L.C.; Ferreira, L.R.; Yamakawa, E.K.; Custodio, E.S.; Aoki, A.R.; Fernandes, T.S.P.; Cardoso, K.H. Centralized and Decentralized Approaches to Demand Response Using Smart Plugs. In Proceedings of the 2014 IEEE PES T&D Conference and Exposition, Chicago, IL, USA, 14–17 April 2014; pp. 1–5. [CrossRef]

12. Shojafar, M.; Cordeschi, N.; Baccarelli, E. Energy-efficient adaptive resource management for real-time vehicular cloud services. *IEEE Trans. Cloud Comput.* **2016**. [CrossRef]
13. Ma, W.J.; Wang, J.; Gupta, V.; Chen, C. Distributed energy management for networked microgrids using online admm with regret. *IEEE Trans. Smart Grid* **2018**, *9*, 847–856. [CrossRef]
14. Liu, N.; Yu, X.; Wang, C.; Li, C.; Ma, L.; Lei, J. An Energy Sharing Model with Price-based Demand Response for Microgrids of Peer-to-Peer Prosumers. *IEEE Trans. Power Syst.* **2017**, *32*, 3569–3583. [CrossRef]
15. Christidis, K.; Devetsikiotis, M. Blockchains and smart contracts for the internet of things. *IEEE Access* **2016**, *4*, 2292–2303. [CrossRef]
16. Liu, H.; Zhang, Y.; Yang, T. Blockchain-enabled security in electric vehicles cloud and edge computing. *IEEE Netw.* **2018**, *32*, 78–83. [CrossRef]
17. Dorri, A.; Kanhere, S.S.; Jurdak, R.; Gauravaram, P. Blockchain for IoT Security and Privacy: The Case Study of a Smart Home. In Proceedings of the IEEE International Conference on Pervasive Computing and Communications Workshops, Kona, HI, USA, 13–17 March 2017; pp. 618–623. [CrossRef]
18. Rios, E.; Prünster, B.; Suzic, B.; Prieto, E.; Notario, N.; Suciu, G.; Ruiz, J.F.; Orue-Echevarria, L.; Rak, M.; Franchetto, N.; et al. *Cloud Technology Options towards Free Flow of Data*; DPSP Cluster: Reggio Calabria, Italy, 2017; pp. 1–110. [CrossRef]
19. Goopal. Digital Point Based on Blockchain Technology. 2015. Available online: http://goopal.online/Z-documents_white_paper.html (accessed on 15 December 2017).
20. Burger, C.; Kuhlmann, A.; Richard, P.; Weinmann, J. *Blockchain in the Energy Transition: A Survey among Decision-Makers in the German Energy Industry*; Deutsche Energie-Agentur GmbH (dena); German Energy Agency: Berlin, Germany, 2016.
21. Kawasmi, E.A.; Arnautovic, E.; Svetinovic, D. Bitcoin-Based Decentralized Carbon Emissions Trading Infrastructure Model. *Syst. Eng.* **2015**, *18*, 115–130. [CrossRef]
22. Shojafar, M.; Canali, C.; Lancellotti, R.; Abawajy, J. Adaptive computing-plus-communication optimization framework for multimedia processing in cloud systems. *IEEE Trans. Cloud Comput.* **2016**. [CrossRef]
23. Pooranian, Z.; Conti, M. RARE: Defeating Side Channels based on Data-Deduplication in Cloud Storage. In Proceedings of the INFOCOM Workshops CCSNA, Honolulu, HI, USA, 15–19 April 2018. [CrossRef]
24. Ping, J.; Chen, S.; Zhang, N.; Yan, Z.; Yao, L. Decentralized transactive mechanism in distribution network based on smart contract. *Proc. CSEE* **2017**, *37*, 3682–3690. [CrossRef]
25. Li, B.; Cao, W.; Qi, B.; Sun, Y.; Guo, N.; Su, Y.; Cui, G. Overview of application of block chain technology in ancillary service market. *Power Syst. Technol.* **2017**, *41*, 736–744. [CrossRef]
26. Ouyang, X.; Zhu, X.; Ye, L.; Yao, J. Preliminary applications of blockchain technique in large consumers direct power trading. *Proc. CSEE* **2017**, *37*, 3737–3745. [CrossRef]
27. Wu, G.; Zeng, B.; Li, R.; Zeng, M. Research on the application of blockchain in the integrated demand response resource transaction. *Proc. CSEE* **2017**, *37*, 3717–3728. [CrossRef]
28. She, W.; Hu, Y.; Yang, X.; Gao, S.; Liu, W. Virtual power plant operation and scheduling model based on energy blockchain network. *Proc. CSEE* **2017**, *37*, 3729–3736. [CrossRef]
29. Tai, X.; Sun, H.; Guo, Q. Electricity transactions and congestion management based on blockchain in energy internet. *Power Syst. Technol.* **2016**, *40*, 3630–3638. [CrossRef]
30. Mattila, J.; Seppälä, T. *Industrial Blockchain Platforms: An Exercise in Use Case Development in the Energy Industry*; Etla Working Papers: Berkeley, CA, USA, 2016.
31. He, Y.; Li, H.; Cheng, X.; Liu, Y.; Yang, C.; Sun, L. A blockchain based truthful incentive mechanism for distributed p2p applications. *IEEE Access* **2018**, *6*, 27324–27335. [CrossRef]
32. Fan, T.; He, Q.; Nie, E.; Chen, S. A study of pricing and trading model of Blockchain & Big data-based Energy-Internet electricity. In Proceedings of the 3rd International Conference on Environmental Science and Material Application (ESMA 2017), Chongqing, China, 25–26 November 2017; pp. 1–12. [CrossRef]
33. Buterin, V. A Next Generation Smart Contract and Decentralized Application Platform. 2014. Available online: https://www.weusecoins.com/assets/pdf/library/Ethereum_white_paper-a_next_generation_smart_contract_and_decentralized_application_platform-vitalik-buterin.pdf (accessed on 25 April 2018).
34. Stark, J. Making Sense of Ethereum's Layer 2 Scaling Solutions: State Channels, Plasma, and Truebit. 2018. Available online: https://medium.com/l4-media/making-sense-of-ethereums-layer-2-scaling-solutions-state-channels-plasma-and-truebit-22cb40dcc2f4 (accessed on 23 July 2018).

35. Warren, W.; Bandeali, A. 0x: An Open Protocol for Decentralized Exchange on the Ethereum Blockchain. 2017. Available online: https://www.0xproject.com/pdfs/0x_white_paper.pdf (accessed on 19 May 2018).
36. Dahlquist, O.; Hagstrom, L. Scaling Blockchain for the Energy Sector. Master's Thesis, University of Uppsala, Uppsala, Sweden, June 2017.
37. Wang, D.; Zhou, J.; Wang, A. Loopring: A Decentralized Token Exchange Protocol. 2018. Available online: https://github.com/Loopring/whitepaper/raw/master/en_whitepaper.pdf (accessed on 2 July 2018).
38. Zhao, C.; He, J.; Cheng, P.; Chen, J. Consensus-based energy management in smart grid with transmission losses and directed communication. *IEEE Trans. Smart Grid* **2017**, *8*, 2049–2061. [CrossRef]
39. Wang, J.; Zhou, N.; Wang, Q.; Wang, P. Electricity direct transaction mode and strategy in microgrid based on blockchain and continuous double auction mechanism. *Proc. CSEE* **2018**. [CrossRef]
40. Nicolaisen, J.; Petrov, V.; Tesfatsion, L. Market power and efficiency in a computational electricity market with discriminatory double-auction pricing. *IEEE Trans. Evol. Comput.* **2001**, *5*, 504–523. [CrossRef]

© 2018 by the authors. Licensee MDPI, Basel, Switzerland. This article is an open access article distributed under the terms and conditions of the Creative Commons Attribution (CC BY) license (http://creativecommons.org/licenses/by/4.0/).

MDPI
St. Alban-Anlage 66
4052 Basel
Switzerland
Tel. +41 61 683 77 34
Fax +41 61 302 89 18
www.mdpi.com

Energies Editorial Office
E-mail: energies@mdpi.com
www.mdpi.com/journal/energies

www.ingramcontent.com/pod-product-compliance
Lightning Source LLC
LaVergne TN
LVHW072000080526
838202LV00064B/6796